TK
6655
.V5
M345
1986
0823649
0 0 3 1 5 0 4 8 8

D0146068

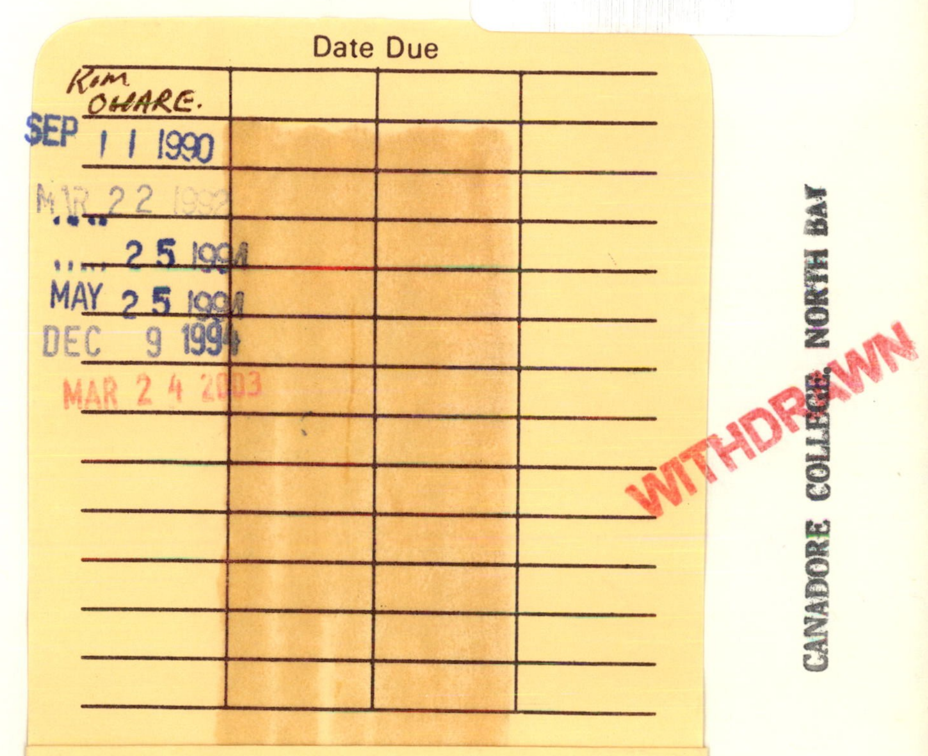
Date Due
Kim OHARE.
SEP 11 1990
MAR 22 1992
25 1994
MAY 25 1994
DEC 9 1994
MAR 24 2003

paul 5928

MAINTENANCE AND REPAIR OF VIDEO CASSETTE RECORDERS

Matthew Mandl

Prentice-Hall, Inc., Englewood Cliffs, NJ 07632

Library of Congress Cataloging in Publication Data

Mandl, Matthew.
Maintenance and repair of video cassette recorders.

Includes index.
1. Video tape recorders and recording—Maintenance and repair. I. Title.
TK6655.V5M345 1986 621.388'332 85-9512
ISBN 0-13-545526-X

Editorial/production supervision and
interior design: Anthony Keating
Cover design: Whitman Studio Inc.
Manufacturing buyer: Rhett Conklin

Printed in the United States of America

10 9 8 7 6 5 4 3 2 1

ISBN 0-13-545526-X 01

Prentice-Hall International (UK) Limited, *London*
Prentice-Hall of Australia Pty. Limited, *Sydney*
Editora Prentice-Hall do Brasil, Ltda., *Rio de Janeiro*
Prentice-Hall Canada Inc., *Toronto*
Prentice-Hall Hispanoamericana, S.A., *Mexico*
Prentice-Hall of India Private Limited, *New Delhi*
Prentice-Hall of Japan, Inc., *Tokyo*
Prentice-Hall of Southeast Asia Pte. Ltd., *Singapore*
Whitehall Books Limited, *Wellington, New Zealand*

Contents

Preface

This text is designed to serve as a guide to the maintenance and repair of video cassette recorders (VCRs). It can be used to advantage by the beginner in electronics because the early chapters provide the necessary groundwork in the makeup and functional aspects of the VCR as well as in the routine maintenance procedures necessary to assure continued operation. The basic technical aspects of circuits and systems are described so that the reader can understand the manner in which picture and sound are recorded. Extensive description of the function of circuits or mathematical equations is avoided whenever possible.

The first four chapters are devoted to a description of the basic signals and systems with a description of the most common problems found in both the VHS and Beta VCR systems. A comprehensive description of the VHS system is given in Chapter 5. This chapter describes the nature of video and sound recording on tape and the particular methodology utilized in the VHS system. The Beta system is covered in Chapter 6 and the essential differences between it and the VHS system are also included.

A thorough discussion of the various types of test equipment is covered in Chapter 7. Though not all of these instruments are essential in 80% of the repairs necessary in VCRs, the inclusion of the various units serves as an excellent reference when the need for understanding a particular type arises. The basic power supply systems are covered in Chapter 8. As with other descriptions of circuitry and mechanical aspects of VCRs the coverage is broad and applicable to all types rather than only to specific VCR units by one manufacturer.

Testing and repair of video and audio sections (including alignment and color circuit adjustments) are covered in Chapter 9. Signal-tracing procedures and parts isolation and replacement are given in Chapter 10 and again the coverage is broad to include all VCR types. The material in this chapter, as well as other chapters, is amply illustrated to clarify essential points and to simplify the test procedures. Certain precautions are emphasized throughout for safety purposes and to minimize damage to the VCR circuitry and components.

A useful Master Index to Common VCR Troubles is given in Chapter 11. Here the various symptoms of faults and defects in VCRs are listed, accompanied by an indication of the probable cause. There is also a reference to the actual section in the book that discusses the troubleshooting and repair procedures necessary for the correction of that fault. This Master Index is comprehensive, since it cross-references virtually all portions of the text material.

Chapter 12 contains various factors relating to certain aspects of the VCR, including the use of a computer to place captions or titles on the tape, stereo factors, camera data, cable TV frequency factors, and a discussion on preemphasis, deemphasis, and the Dolby system. The Appendix includes TV station allocations, cable TV frequencies, television standards in broadcasting, color television reference standards, and component color codes.

Matthew Mandl

1

Basic VCR Signals and Systems

1-1. BASIC SIGNALS

Video cassette recorders (VCRs) operate under two distinct systems: the Beta, originally developed by Sony Corporation, and the VHS (Video Home System), introduced by JVC (Japan Victor Co.). A number of manufacturers utilize the Beta, while others employ the VHS method. The systems are not compatible either in the recording or playback mode. Each system uses a different method for recording the signals, and consequently a tape recorded on the Beta system cannot be played on the VHS, nor can a VHS recording be played on a Beta VCR. The cassette housings differ also, with the VHS units being somewhat larger, as more fully discussed later in this chapter. (The circuitry and technical recording aspects of both the VHS and the Beta system are detailed in Chapters 5 and 6.)

Although the recording procedures differ, each system must process the *same type of signal* received from the television antenna or cable system. Most maintenance procedures can be performed without a knowledge of the type of signals encountered. Similarly, many repairs can be made without knowledge of the nature of the signals processed. For advanced servicing procedures, however, the more involved repairs are expedited considerably if one is familiar with the nature of the signals encountered and the manner in which they are processed. Hence, the basic signal systems are given in this section, and their utilization in terms of television and sound recording and reproduction are covered in a later section.

All signals utilized in video and sound recording and reproduction are alternating-current types. As such they closely resemble the alternating-current signals that make up the electricity in our power mains. For the latter, however, a single-frequency signal is utilized, while in audio and high-frequency applications the signals may consist of various frequency types that also vary in amplitude. The ac signal in our power mains has a fairly constant amplitude and a fixed-frequency rate of 60 hertz (Hz), which means that one complete waveform (cycle) occurs 60 times per second.

The basic ac signal waveform is shown in Fig. 1-1(a). Note that the voltage (or current if that were being designated) increases from zero value to a positive peak level and back to zero. Next there is a gradual increase to a negative peak value and again a decline to zero. One segment of incline and decline represents one-half of the cycle and is designated as an *alternation*. As shown, there are two alternations per cycle. A pure sine wave has equal amplitude alternations, an unbroken incline and decline of amplitude, and equal time durations for successive alternations.

As shown in Fig. 1-1(b), the peak voltage (or current) is termed the *peak value*, and at the 0.707 point the designation is *average value* or the *root-mean-square* (rms) value. The latter is the amount of electric energy required to perform the same work as direct voltage and current. The formations shown in Fig. 1-1 are termed *sine waves*, and such sig-

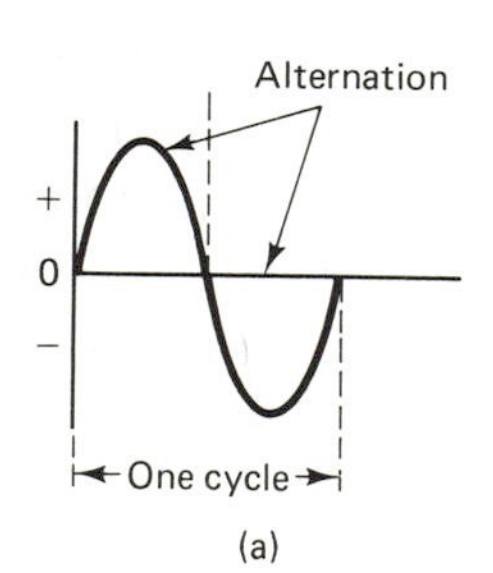

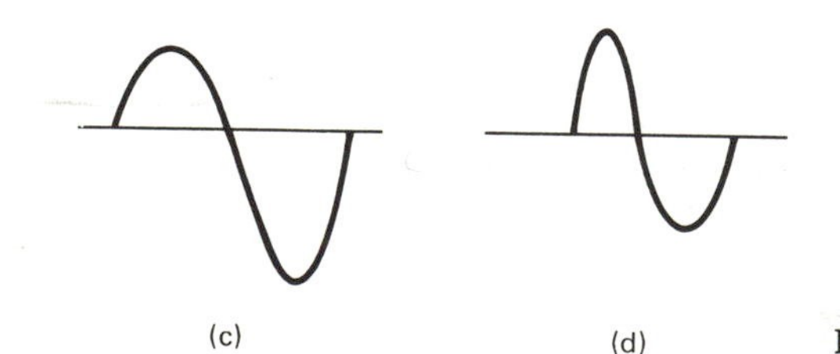

Figure 1-1 The ac cycle.

nals have a wide range of frequencies in audio and television systems. Thus, such signals of various frequencies and amplitudes are the foundation for speech signals, music signals, video signals, noise signals, and so on.

A pure sine-wave signal has only one particular frequency, but when such a signal is distorted, it contains signals of higher frequencies. Such additional signals are usually termed *harmonics*, and they have successively higher frequencies above the fundamental frequency. One common type of distortion is that caused by amplitude differences such as shown in Fig. 1-1(c). Here the first alternation has a lower amplitude than the second, and thus the signal is distorted in comparison to that shown at Fig. 1-1(a). Another type of distortion is shown in Fig. 1-1(d) where the duration of one alternation differs from the other. For both types of distortion harmonic signals are generated. (See also Sec. 7-5.)

The basic sine-wave signals are often modified to produce various other signal types. These comprise the signals that are routed through various circuits for amplification or other processing as required by video recorders. To perform such circuit functions, direct-current (dc) power must be applied to the transistor circuits or chips. Such dc power can be obtained from batteries or power supplies (Chapter 8). Power supplies convert the alternating current from the power mains into the necessary direct current for circuit operation. It must be remembered that direct current so utilized is not considered a signal.

Various waveshapes are shown in Fig. 1-2. The pure sine-wave signal is shown in Fig. 1-2(a) for comparison to the others. In Fig. 1-2(b) we have a succession of single-polarity alternations of the type obtained by half-wave power supply systems. These pulses are filtered into pure direct current. In Fig. 1-2(c) are closely spaced alternations produced by full-wave rectification as detailed in Chapter 8. In Fig. 1-2(d) is shown a square-wave signal train. Such square waves have positive and negative polarities identical to the sine wave. Square waves are useful for test purposes, as described more fully in Chapter 7. Similarly, the single-polarity pulse shown in Fig. 1-2(e) is useful for test purposes. In addition, however, such pulses are used in the television picture signal for synchronization and blanking purposes, as described more fully later in this chapter.

A sawtooth waveform is pictured in Fig. 1-2(f). Such an incline and abrupt decline produce a signal that is useful in sweeping picture information across a television screen or for moving signal data across the screen of a cathode-ray oscilloscope. (See Chapter 7.) The waveform in Fig. 1-2(g) is a distorted square-wave type wherein some of the high-frequency harmonic components have been lost. (If all the harmonic components are eliminated, the square wave becomes a sine

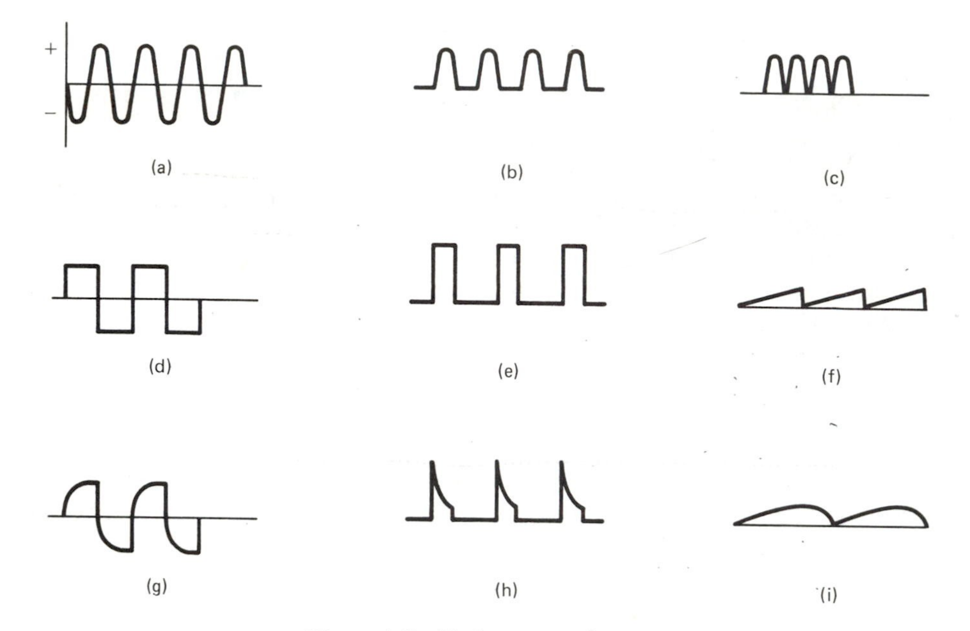

Figure 1-2 Various waveshapes.

wave.) The signal spikes shown in Fig. 1-2(h) are useful for triggering purposes where a circuit must initiate some process very quickly. All such waveforms are encountered in video systems.

1-2. AUDIO AND VIDEO SIGNALS

The representation of signal waveforms does not identify the frequency. Compare, for instance, the sine-wave signal of Fig. 1-2(a) with those shown in Fig. 1-3. Although either representation could be an audio signal or a high-frequency television signal, generally the waveform in Fig. 1-2 designates the lower-frequency audio signal, while that of Fig. 1-3 would indicate a radio-frequency (RF) type of signal such as used in television and radio broadcasting. Again, for the waveforms shown in Fig. 1-3 we could be representing a radio signal having a frequency of 900 Hz, or it could represent a video picture signal having a frequency of 77.25 MHz (77,250,000 Hz). In either case we can display a signal having a constant frequency and amplitude as shown at the left of Fig. 1-3 or one having an amplitude that changes constantly as shown at the right. Audio and video signals are characterized by having varying amplitude and frequency changes. A picturization is only representative since it would be impossible for us to draw the thousands or millions of individual sine-wave signals of various frequencies and amplitudes. In-

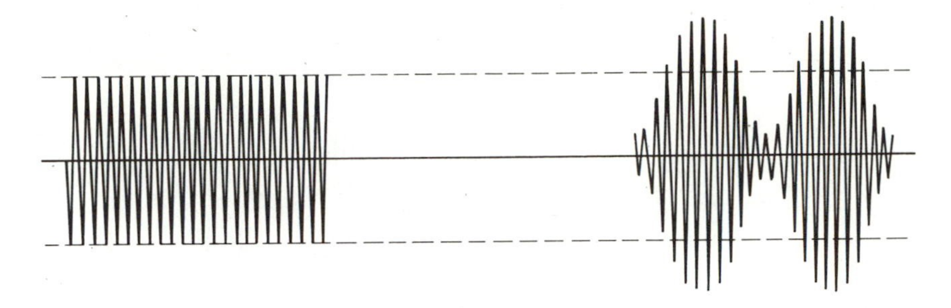

Figure 1-3 RF signal representations.

stead, the illustrations are generalizations and are understood to have the required number of alternations.

Most audio signals lie within the range of 50 to 15,000 Hz, though harmonics will extend to 20,000 Hz or more. Table 1-1 gives some representative ranges of sound. Transmitted video signals, however, have a frequency of 55.25 MHz for channel 2 and range up to 759.25 for channel 6 (see Appendix A3). Sound and picture information is sent through space by utilizing a much higher-frequency signal to carry the lower-frequency signal information. As might be expected, the higher-frequency signal is thus called a *carrier*, though actually it has a characteristic which, at the receiving end, permits the *reproduction* of the audio signal rather than carrying it through space in its original form.

In television transmission there are two carriers. One carrier contains the picture-signal information and the other the sound information. Two processes are involved: *amplitude* modulation and *frequency* modulation. A brief description of these processes will make clear the nature of the signals received by a VCR and a television set. The basics of amplitude modulation are illustrated in Fig. 1-4. The carrier signal generated by the circuits has a constant amplitude when there is no

TABLE 1-1 Typical Frequency Ranges of Various Sound Signals

Sound Type	Approximate Frequency Range (Hz)
Desirable range for good speech intelligibility	300–4000
Audibility range (normal hearing, young person)	16–20,000
Piano	26–4000
Baritone	100–375
Tenor	125–475
Soprano	225–675
Cello	64–650
Violin	192–3000
Piccolo	512–4600
Harmonics of sound	32–20,000

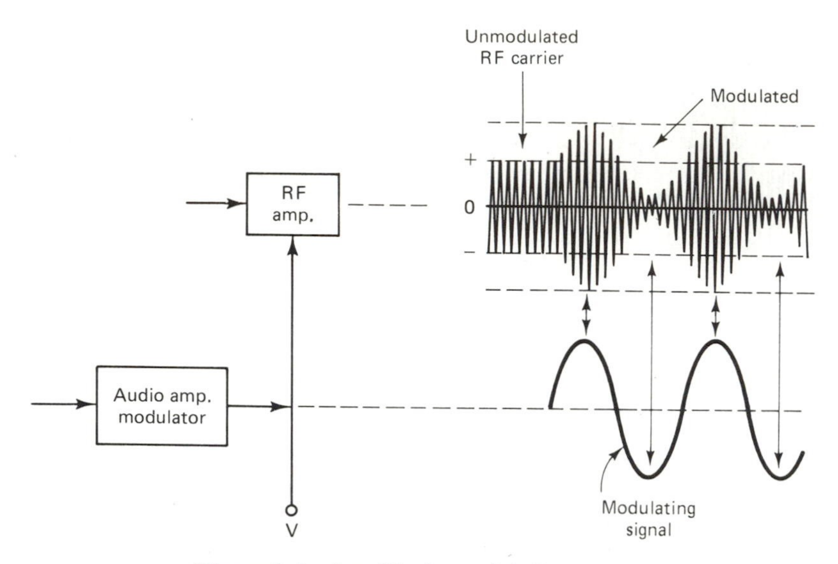

Figure 1-4 Amplitude modulation process.

modulation involved. When audio signals are introduced, the resultant carrier waveform increases and decreases in amplitude in accordance with the rise and fall of the modulating audio signal.

As shown, the audio signal can be fed in series with the dc applied to the carrier circuits. It adds and subtracts from the level of power being fed to the carrier circuitry. Thus, the amplitude changes of the carrier produces amplitude modulation. Since the latter represents a distortion of the carrier amplitude, it produces additional signals of other frequencies. These new signals are termed sidebands. If a 1000-kHz carrier is modulated by a 2-kHz audio signal, two sidebands are created, one having a frequency of 1002 kHz and the other 998 kHz. (The lowercase k represents 1000.) For a 5-kHz audio tone the sidebands produced would be 1005 and 995 kHz. Amplitude modulation is not only used in radio broadcasting but also for transmitting the video signals in television.

The amplitude changes in an amplitude-modulated carrier signal represent the *composite* waveform (made up of the carrier plus the sidebands). For a given audio modulating signal, the individual carrier and sideband signals have a constant amplitude. When, however, all are combined, the composite RF signal exhibits amplitude changes.

Another process for modifying a carrier is that called *frequency modulation.* In the latter the carrier frequency is shifted above and below its normal frequency at a rate established by the frequency of the audio signal. The formation of the resultant carrier is illustrated in Fig. 1-5. A number of sidebands are generated, but only a few clustered

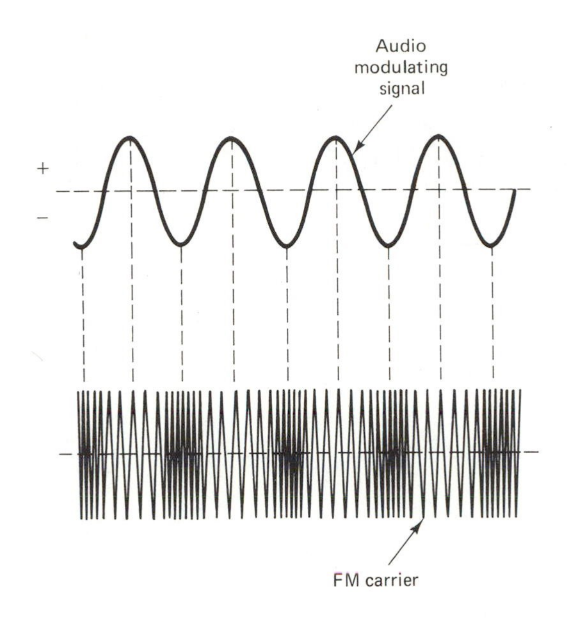

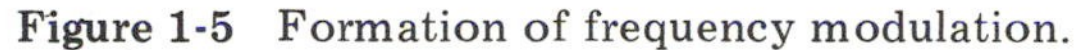

Figure 1-5 Formation of frequency modulation.

around the carrier frequency are of significance in transmission and reception. In the frequency-modulation process the extent of the carrier deviation determines modulating signal amplitude. The degree of sideband spacings from the carrier is set by the frequency of the modulating signal. In television transmission the audio portion is transmitted using frequency modulation.

When either AM or FM signals are detected by appropriate circuitry in the television receiver, the original modulating signals are produced. For the picture signals the detected signal consists of a complex waveform made up of video, blanking, and timing information. For a video signal to reproduce sharp images, an enormously wide band of frequencies is necessary because the picture signals extend to 4 megahertz (MHz). One segment of the signal may appear as shown in Fig. 1-6. Here the initial portion at the left consists of the video signal information, while the waveform at the right represents a horizontal blanking pulse and a horizontal synchronization pulse mounted on top. Such a waveform is essential because the signal must trace picture information across the screen from left to right and must also retrace downward from right to left as it travels down the picture tube face. Whenever it finishes a scan across the screen, the latter is momentarily blanked out, and a synchronizing pulse initiates the beam retrace to the left, as shown in Fig. 1-7. When the beam scan reaches the bottom of the picture tube, a complete field has been displayed. (See also Appendix C and Table C-1.) A series of pulses is used to cause the beam to go back

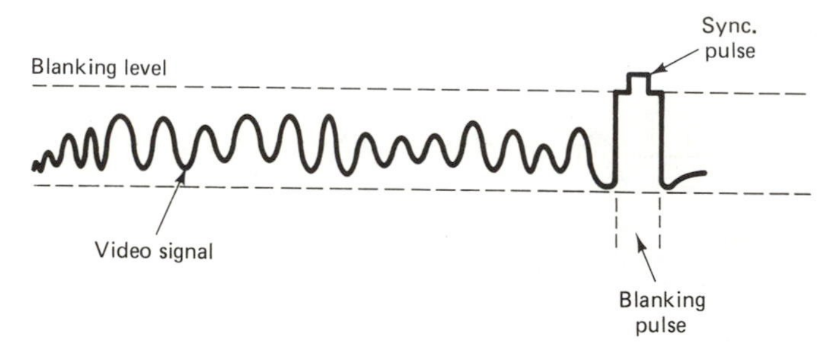

Figure 1-6 Video and pulse signals.

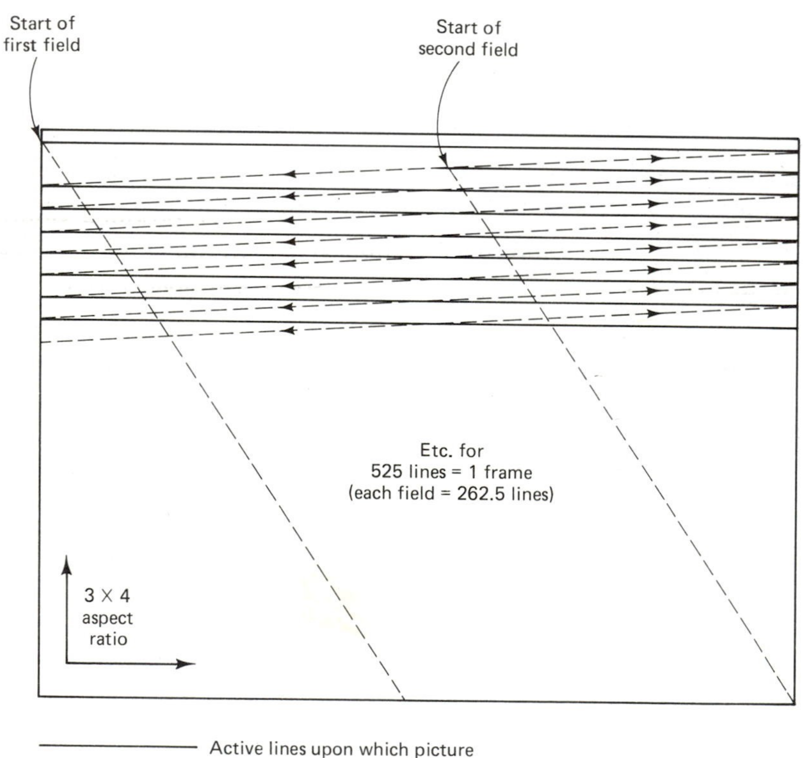

Active lines upon which picture information is traced

Retrace lines of beam (blanked out during picture reception)

Figure 1-7 Interlaced video scanning.

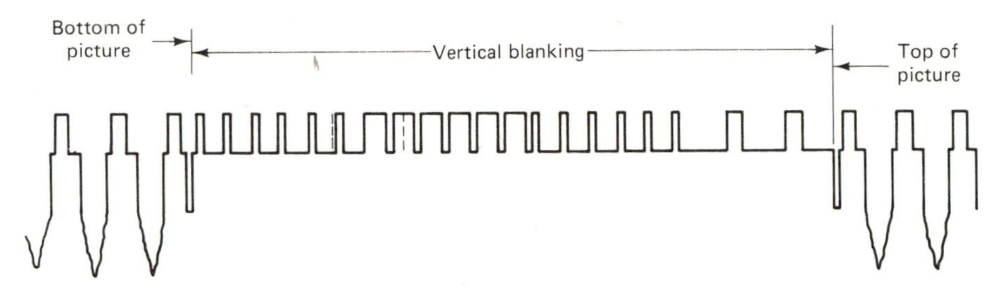

Figure 1-8 Vertical blanking waveforms.

to the top of the picture tube and start the horizontal scanning again. For the vertical retrace process, a series of pulses, as shown in Fig. 1-8, is used. This is a combination of pulses that not only triggers the beam to sweep back to the top of the screen but also maintains the horizontal circuits in perfect synchronization. Two complete fields constitute a frame. There are 262.5 scan lines per field and 525 lines per frame. The two fields that constitute a frame are interwoven (termed *interlace*) to increase picture detail and decrease flutter. The complete line trace on the screen (a white screen minus a picture) is termed a *raster*. Good interlace is obtained by careful adjustment of the vertical hold control in a TV set or monitor.

Since video signals that range up to 4 MHz produce sidebands spaced 4 MHz above and below the carrier frequency, such signals would utilize a prohibitive amount of spectrum space. If each television station required an 8-MHz bandpass, the number of television stations would have to be reduced. Thus, to conserve spectrum space, most of the lower-frequency sidebands are eliminated with only a remnant group (called *vestigial sidebands*) remaining, as shown in Fig. 1-9. Even with this omission of lower sidebands, however, the combined video and audio signals require a 6-MHz bandpass. This is still a very

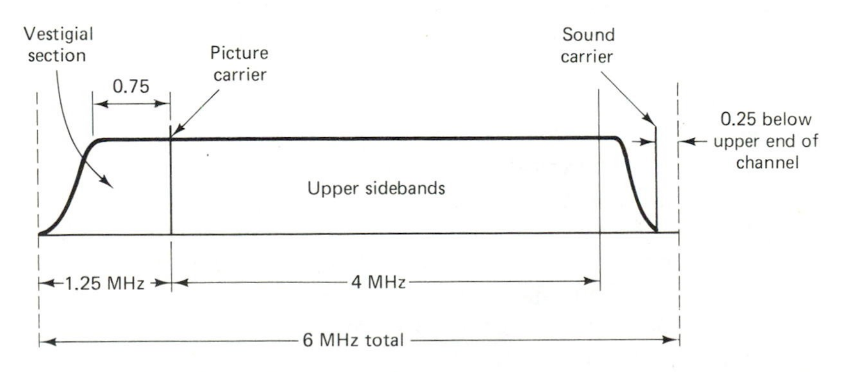

Figure 1-9 Video bandpass after most lower sidebands suppressed.

wide bandpass, and therefore it is essential that the circuits of video cassette recorders and television receivers are designed and maintained at this bandpass for proper picture rendition.

1-3. SIGNAL MIXING

A video cassette recorder receives television signals from an antenna or a cable system and processes these signals for tape recording purposes. The VCR is also capable of feeding the TV signals it receives directly to a television receiver. The VCR input and output systems are so designed that the incoming signal can be processed in the following manner:

1. An incoming television signal can be fed directly to the TV receiver for viewing without recording.
2. An incoming signal can be recorded in the VCR and viewed on the TV receiver at the same time.
3. An incoming signal can be viewed on the receiver at the same time that the VCR is tuned to a different signal and records the latter.
4. An incoming signal can be recorded with the television receiver shut off (a process utilized when the timer is set to record a program sometime in the future).

To accomplish such various operational modes, the VCR's internal circuitry must have switching capabilities and must be linked properly to a television receiver by a coaxial cable. Such coaxial cable linkages among the antenna, the VCR, and the television receiver are covered in Chapter 2. As with television receivers, the VCR is equipped with a tuner circuit for proper selection of the television stations received from the antenna or the cable. In the tuners of both the VCR and television receiver the incoming signals undergo a mixing process. An understanding of this function will aid in diagnosing troubles that may occur in these sections.

Television and radio receivers utilize the *superheterodyne* principle. When a station is received by a television set, for instance, the tuner circuits mix the incoming television signal with an extra signal generated by one of the circuits in the tuner. This mixing process is also termed *heterodyning*, and it creates a new signal that contains all the information of the original signal but has a different frequency. This new frequency signal is called the *intermediate-frequency* (IF) signal, and it always has the same predetermined frequency regardless of which station is tuned in. As new stations are tuned in, the tuner oscillator (the signal generator) keeps in step with the incoming frequency signal to produce the identical intermediate frequency. Thus, IF amplifiers that follow the tuner only have to handle signals of the intermediate frequency and hence can be designed for proper bandpass and optimum gain.

In radio we are only concerned with one carrier, but in television we encounter an amplitude-modulated video carrier and a frequency-modulated sound carrier. Consequently two IF signals are obtained, one for the video and one for the audio. Frequencies which have become standard for video IF stages are 45.75 MHz for the video carrier and 41.25 MHz for the sound carrier. The mixing process and signal frequencies for channel 9 are shown in Fig. 1-10(a). Note the nearness of adjacent-channel signals consisting of the sound carrier for channel 8 and the picture carrier for channel 10. When the tuner receives channel 9 in our example, the adjacent-channel signals may also ride in to cause interference, particularly under certain atmospheric conditions. Hence, traps are employed to minimize such interference, as discussed later and illustrated in Fig. 1-13.

For channel 9 the oscillator in the tuner would be generating a signal of 233 MHz. When this signal mixes with the picture carrier signal of channel 9, we obtain 45.75 MHz:

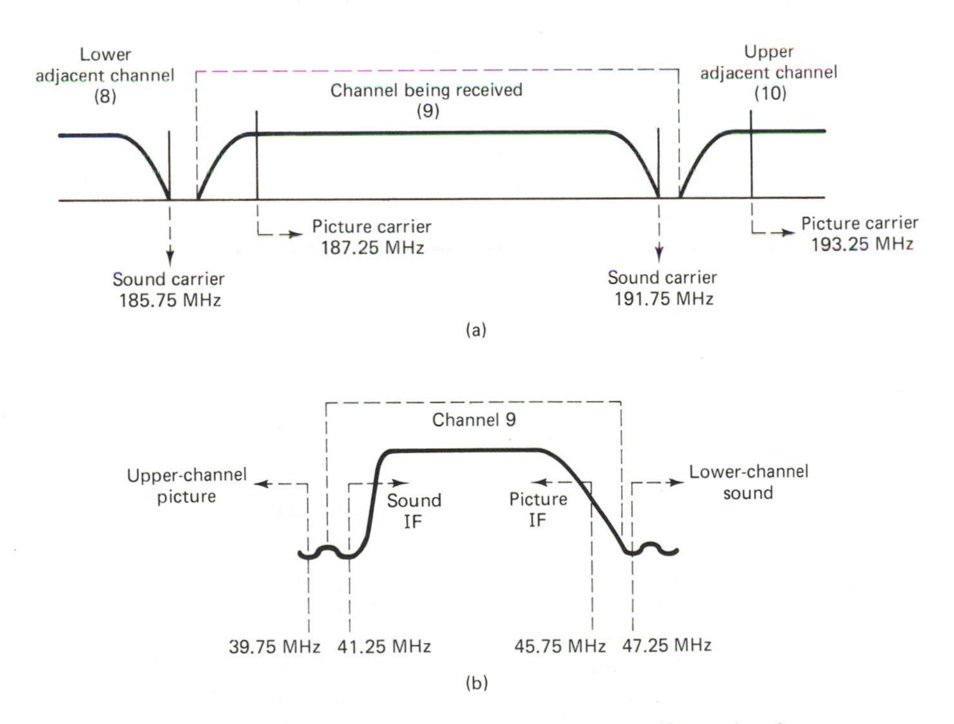

Figure 1-10 Mixing of station and tuner oscillator signals.

$$\begin{array}{rl} 233.00 & \text{(oscillator signal)} \\ -\underline{187.25} & \text{(channel 9 picture carrier)} \\ 45.75 & \text{MHz} \end{array}$$

When the channel 9 sound carrier is mixed with the oscillator signal (233 MHz), we obtain the sound IF signal with a frequency of 41.25:

$$\begin{array}{rl} 233.00 & \text{(oscillator signal)} \\ -\underline{191.75} & \text{(channel 9 sound carrier)} \\ 41.25 & \text{MHz} \end{array}$$

Some of the lower-adjacent-channel sound carrier signal may also ride into the tuner and mix with the 233-MHz oscillator signal to produce a signal having a frequency of 47.25 MHz. This frequency is near enough to the IF bandpass frequencies to cause interference:

$$\begin{array}{rl} 233.00 & \text{(oscillator signal)} \\ -\underline{185.75} & \text{(channel 8 sound carrier)} \\ 47.25 & \text{MHz} \end{array}$$

Some upper-channel picture carrier may also enter the tuner and cause interference, producing a frequency of 39.75 MHz:

$$\begin{array}{rl} 233.00 & \text{(oscillator signal)} \\ -\underline{193.25} & \text{(channel 10 picture carrier)} \\ 39.75 & \text{MHz} \end{array}$$

Thus, the signal-mixing process in the tuner produces the IF signals described, and the resultant IF bandpass that is required is shown in Fig. 1-10(b). Note that now the picture and sound signals have been transposed from left to right and right to left. Note also the dips in the bandpass where the traps in the tuner minimize signal sensitivity at these frequencies. For best reception and maximum picture detail, the bandpass response of the intermediate-frequency amplifiers must be as shown in Fig 1-10(b). Most IF stages in modern VCRs, television receivers, and monitors have stabilizing frequency-control ceramic or crystal filters, and hence little trouble is experienced with the detuning of the IF stages. If they appear to be operating poorly, other checks and tests should be made, as detailed more fully later.

The same frequencies shown at the bottom of Fig. 1-10(b) are obtained regardless of the station tuned in, as mentioned earlier. The oscillator frequency would change for each station so that the same intermediate-frequency signals and adjacent-channel signals are obtained. The adjacent-channel signals are diminished sufficiently so they are of little consequence. The sound and picture IF signals are then fed to IF amplifier circuits and finally applied to detectors for deriving the picture information (that is sent to the picture tube in receivers) and the audio information (that is applied to the speaker in receivers). In VCRs the video and audio signals are also applied to output jacks, as more fully described in Chapter 2.

A video cassette recorder also has a tuner similar to those in television receivers to permit the selection of incoming stations. The VCR must, however, also be designed to produce a television signal having a standard channel frequency. Thus, in playing recorded video tape or when using the VCR as the primary tuner (instead of the tuner on the television receiver), the VCR must produce signals that are either equivalent to the broadcast frequencies of channel 3 or channel 4. A switch on the rear of the VCR selects a channel 3 or a channel 4 output, depending on which channel is clear for the particular receiving area. Thus, if a local station is on channel 3, the VCR is set for channel 4. The television receiver must then be set for channel 4 also. If channel 4 is in the locality, the VCR switch is set for channel 3, and the receiver is then tuned to the same station. This procedure minimizes the interference which might occur if the VCR only had an output for channel 3, for instance. Local signals could be sufficiently strong and cause interference.

The VCR tuner can be used to select the station to be viewed in lieu of using the TV receiver's station selector. The VCR's tuner is also utilized when viewing recorded material from the VCR tapes. The recorded material is read by the VCR head and converted by the VCR tuner to either channel 3 or 4, as mentioned earlier.

1-4. REDUCING INTERFERENCE

Because the video and sound carriers as well as the IF signals are of such a high frequency, precautions must always be taken to minimize signal loss in both the VCR and television circuit. When high-frequency signals are involved, signal loss can be exceptionally high if proper precautions are not observed. A poor linkage system between the VCR and the television receiver can cause sufficient signal loss to cause the picture quality to suffer materially. A poor installation may also cause interference in the picture because of spurious signals that may ride in with the desired signal. A major cause is improper shielding or a disturbance of the shielding originally established. High-frequency signals are more easily shunted than low-frequency signals and hence diminish the desired signal. *Capacitance* effects between signal wires and the chassis (or adjacent wires) can contribute to signal loss. Capacitance is a condition set up between any two metal objects. Capacitance has the ability to shunt or transfer signals having ac characteristics. The higher the frequency of the signal, the greater the shunting effect and hence the more serious the loss that occurs.

Electrical sparks or the electrical interference generated by poorly designed motors may cause interference because of radiated high-frequency noise signals that are generated. The nearer such interference is to the area of reception, the greater the chance of being picked up by any wires carrying the video signal. Interference is likely when proper shielding has been altered or removed or when linkages are made with twin-lead transmission lines instead of coaxial cable. The latter is a self-shielding transmission line because it has an outer metal braid which is placed at ground potential and which acts as a shield for the inner wire for transfer of high-frequency signals. More detailed discussions on this subject will be found in Chapter 2.

1-5. TUNER CIRCUITS

As the name implies, the tuner circuit is utilized in both the VCR and television receivers to select a specific television station. In modern VCR and TV systems the mechanical knob-turning method of station selection has been replaced by the all-electronic systems. Thus, the problems of noisy and otherwise defective tuners created by constant station-selection switching of the mechanical tuners have been eliminated. Instead of the variable capacitors which tune by meshing groups of plates within each other the present-day systems use varactor diodes that undergo a change of capacitance when the applied voltage is altered. Thus, such diodes can replace the mechanical capacitors.

In varactor tuning systems the station-selector section can employ a series of push buttons, as shown in Fig. 1-11. Here each station is tuned in by adjustment of the variable resistors (R_1, R_2, and so on) that may be used. For instance, diode D_1 is the varactor diode which tunes to resonance with inductor L_3. The initial tuning is adjusted by resistor R_1, and thereafter the push-button switch would close the circuit for selection of a particular channel. The next channel selected is tuned by resistor R_2 and so on for the entire section. The push buttons can be replaced by electronic sensors for touch control, though this system would require individual circuits.

The basic circuitry of a VHF tuner is shown in Fig. 1-12(a). An RF amplifier increases the level of the incoming signal (obtained from the antenna) to the level needed for mixing with the signal generated by the tuner oscillator. As mentioned earlier, the resultant intermediate-frequency signals are then amplified by the circuits discussed in Sec. 1-6 and subsequently detected by the demodulators covered in Sec. 1-7.

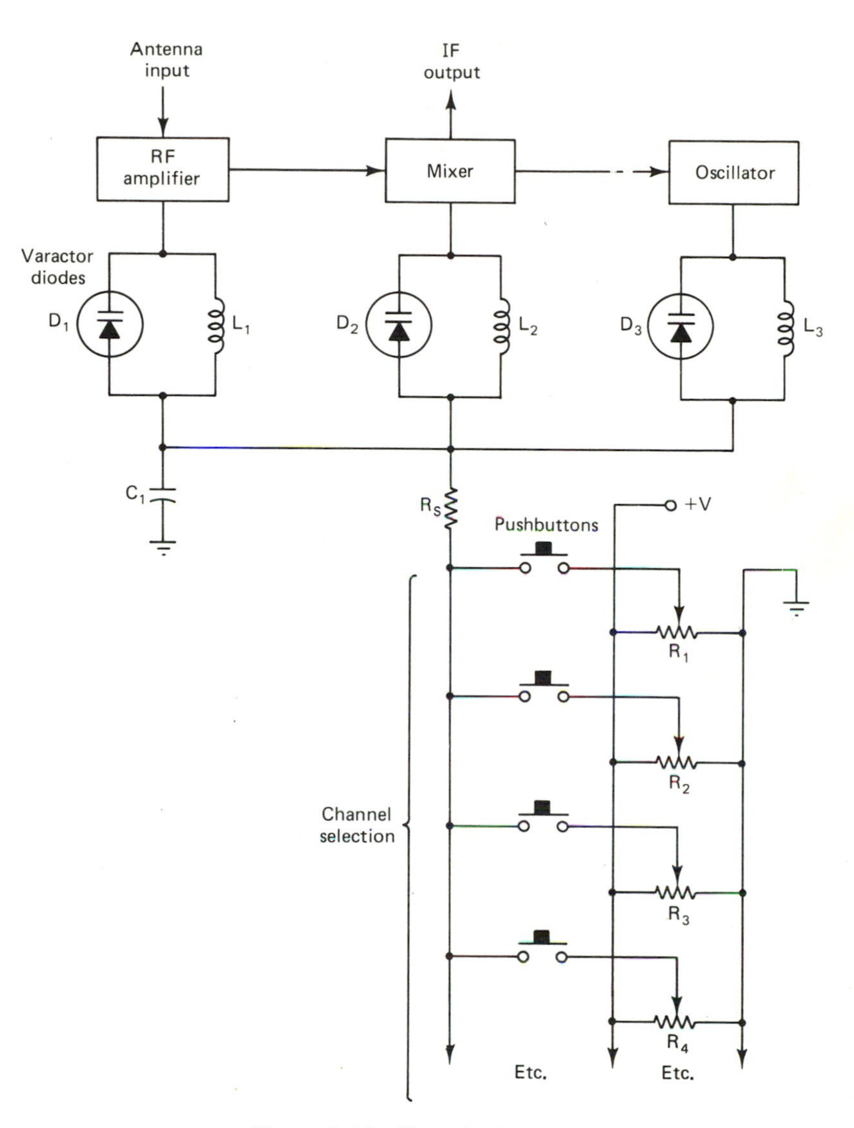

Figure 1-11 Varactor tuner.

For the circuits shown, conventional *npn* transistors are used. Both the incoming RF signal and the signal generated by the oscillator are coupled to the mixer input (Y) where the heterodyning process occurs. At the collector element output of the mixer a tuned resonant transformer section transfers the signal to the IF stages.

Input lines are present for automatic gain control (AGC) voltages obtained from the detector system. The AGC input tends to stabilize the gain of any channel selected and minimizes signal fading. Similarly,

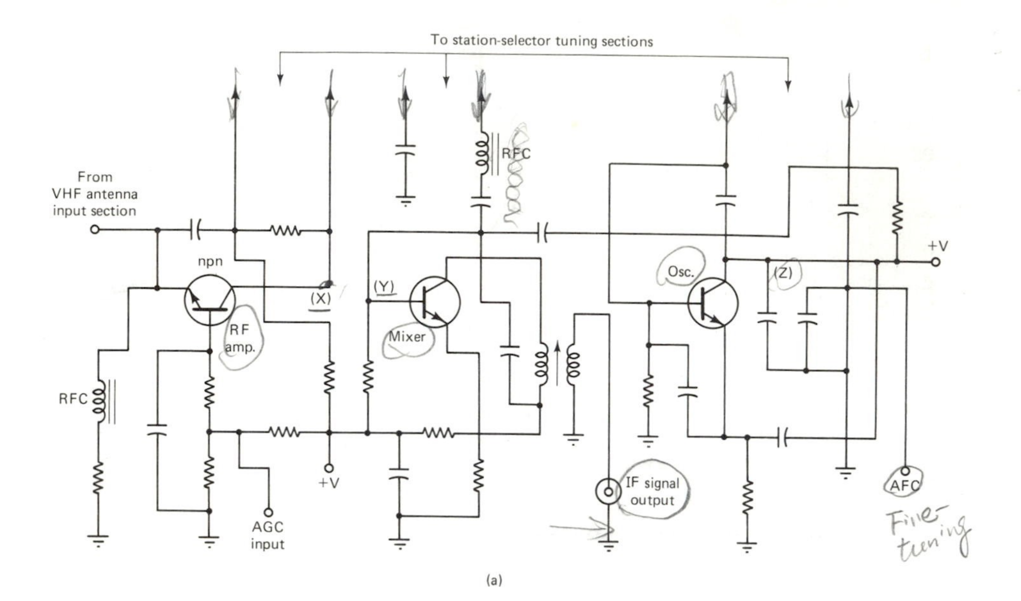

(a)

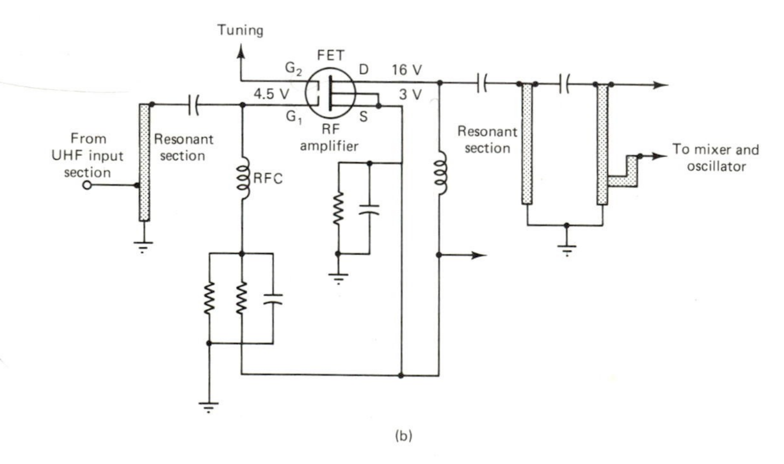

(b)

Figure 1-12 Tuner circuits.

the AFC input at the right regulates the frequency control of the station tuned in and provides for automatic regulation of the fine-tuning process for the station selected. The tuning for the system shown can be of the type shown in Fig. 1-11 or can consist of electronic touch tuning. Test points are sometimes provided for easy checking of voltages and signals. Test point (*X*), for instance, provides a terminal easily accessible for testing the voltage present at the collector of the RF amplifier transistor. Test point (Y) provides for a reading of the input signals to make sure both the RF signal from the amplifier as well as that from the oscillator are present. Test point (Z) is for testing the applied voltage present at the collector of the oscillator transistor.

For UHF station selection a separate tuner section must be utilized because of the much higher signal frequencies involved. In such a tuner special high-frequency resonant sections are employed, as shown in Fig. 1-12(b). These resonant sections consist of short lengths of coaxial cables or either parallel wires or metal rods. In the ultra-high-frequency or microwave regions such sections exhibit significant characteristics of inductance and capacitance and hence are extensively used to form the resonant circuits of amplifiers, oscillators, or mixer circuits.

In Fig. 1-12(b) a field-effect transistor (FET) is shown for the UHF amplifier. The input signal from the antenna input section is coupled to the resonant section at a point above the ground connection that provides for an impedance match between the antenna system and the gate (G_1) input to the RF amplifier. The signal is transferred to the gate from the top of the resonant section as shown. The amplifier output signal is obtained from the drain (D) element of the FET and transferred to a parallel-rod section of resonant line as shown. An inductive loop coupling to one of the resonant elements transfers the output signal to the mixer stage. The mixer and oscillator stages are typical circuits that differ from the VHF types only in the added UHF precautions utilized and the use of resonant sections.

In schematics as shown in Fig. 1-12(b) the voltages present at the elements are usually given for test purposes. Thus, at the drain (D) element we would expect 16 V under normal operating conditions, 3 V at the source (S) element, and 4.5 V at the second gate (G_2) element. Tuning circuits are interconnected with the combined VHF and UHF tuning sections using push buttons, touch tuning, remote control, or, in older receivers, the rotary-switch mechanical system.

The entire tuner assembly is mounted in a small metal container that shields the signal circuits from interfering signals that may be present in the area. Usually when defects occur in such tuners, the repairs are difficult because of the need for high-frequency precautions that require precise replacement and tuning techniques to assure satisfactory performance. In many instances defects in either the UHF or VHF tuners are corrected by replacing the entire tuner section. To facilitate replacement, such sections are often easily disconnected from the chassis and readily removed from the chassis circuitry because of plug-in connecting wires and cables. Trouble localization factors are covered in later chapters. (See also Fig. 10-6.)

1-6. THE IF SYSTEM

In tuning to any particular station, the frequency of the oscillator's signal is automatically altered so that in mixing with the incoming television signal an IF signal is created that has a fixed frequency. This IF signal frequency thus remains the same regardless of the station tuned in. Because of the single frequency, the IF stages only need an initial tuning at the factory and rarely require additional tuning thereafter. If it appears that the tuning is incorrect, the circuit should first be checked for defective components that may be causing such symptoms. If tuning adjusting units are readily accessible, it is tempting to try retuning to see if this corrects the problem. If this is done, extreme care must be taken not to disturb the tuning of the IF stages to the degree where poor reception occurs if a defective component should later be found and replaced. It is a good policy to leave tuning undisturbed until all tests and checks have finally indicated that tuning is necessary.

The IF amplifiers are conventional transistorized circuits with special components added to process the high-frequency television signals. A typical system of this type is shown in Fig. 1-13(a). Note that the output from the VHF-UHF tuner sections is connected to three series circuits (to ground) composed of capacitors and inductors. These are resonant traps that are tuned to a specific signal frequency. A series circuit of this type provides a shunt path for undesired signals and thus prevents their entry to the IF system. One trap is tuned to 39.75 MHz and thus shunts most of the interference that would be caused by an adjacent-channel picture carrier. Another trap is tuned to 47.25 MHz to diminish any interference which may be caused by the sound signal of an adjacent channel. The last trap is tuned to 41.25 MHz, which is the VCR's or TV's own sound IF signal received from the tuner. This signal is diminished sufficiently to prevent possible interference with the picture signal. Since these trap frequencies remain the same regardless of the frequency of the station to which the VCR or television is tuned, they rarely need retuning once they have been set to minimize interference. (See also Fig. 9-5.)

As shown in Fig. 1-13(a), the coils of the traps have adjustable tuning cores (represented by an arrow beside the coil). Often ceramic filters (C_f) are used in IF stages to aid in maintaining a fixed bandpass characteristic. A typical circuit of this type is shown in Fig. 1-13(b) and is superior to the older capacitor-inductor systems. Crystal filters are also used instead of the ceramic filters to maintain circuit tuning stability and to provide better rejection of undesired signals that would interfere with the picture or sound.

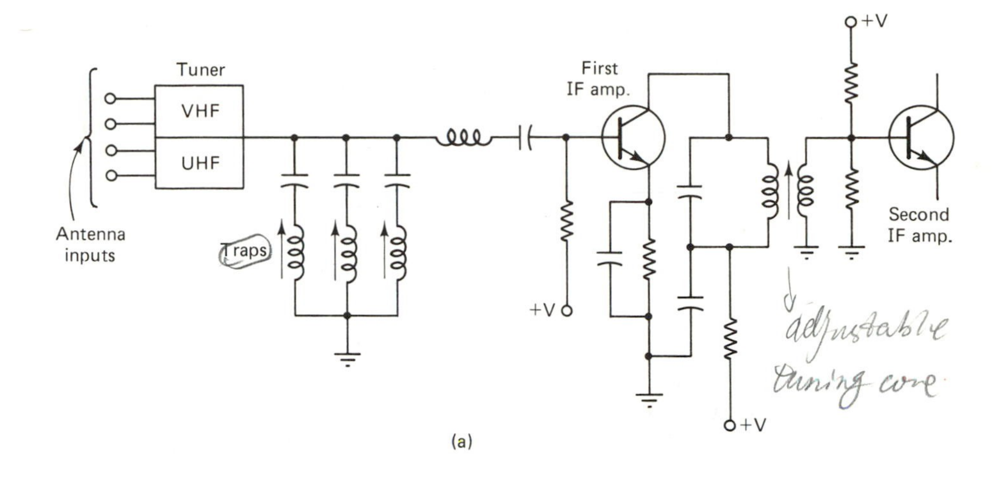

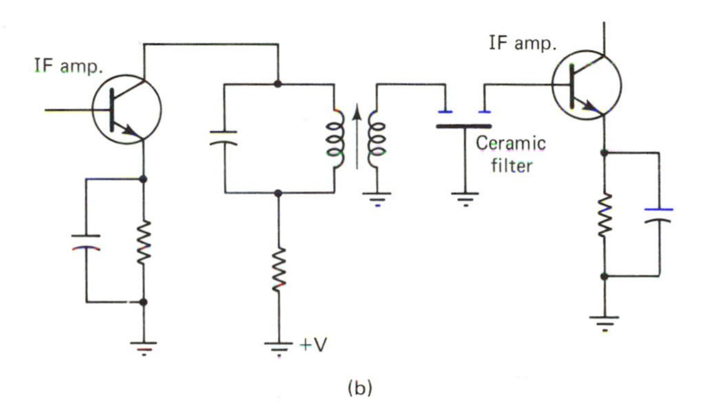

Figure 1-13 IF circuits.

1-7. INTEGRATED CIRCUITS

Integrated circuits (ICs) are also utilized for IF stages and, depending on design, may contain several amplifier sections. Although such integrated circuits have been manufactured in circular forms, the rectangular shaped unit is extensively used. As shown in Fig. 1-14(a), the integrated circuit unit has a number of prongs (some units containing two or three dozen). These prongs are bent down from the flat section and fit into matching sockets. The latter are usually soldered into a printed-circuit board. Such IC units may have an indentation at one end to localize the start of the prong-numbering system. In other instances an unbroken rectangular form was used with an imprinted dot at one end for identification.

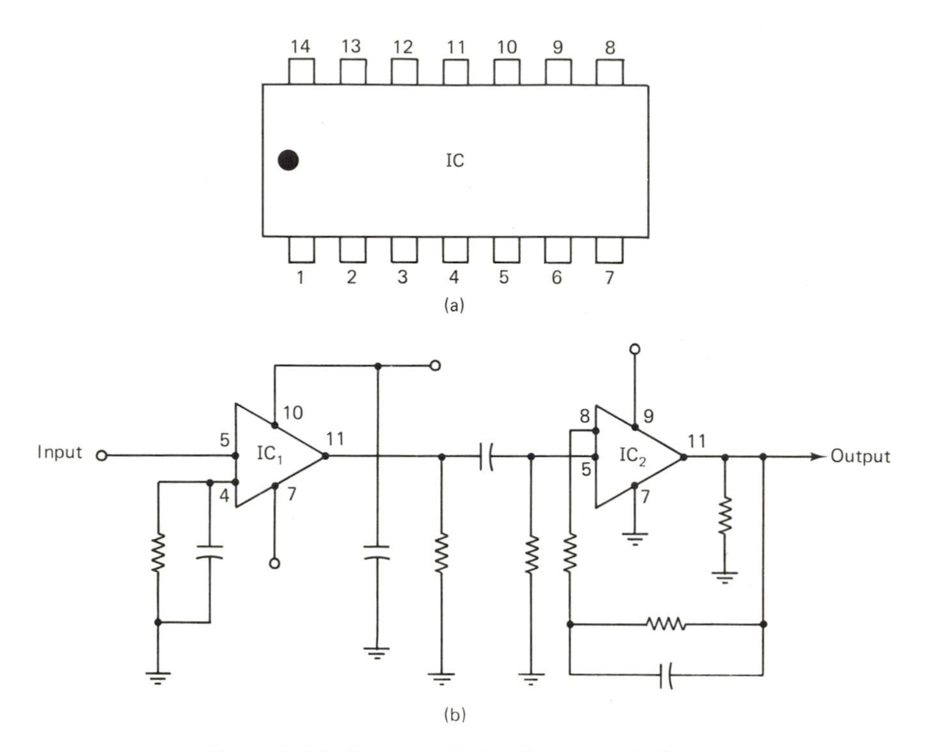

Figure 1-14 Integrated-circuit representations.

Schematically, the integrated circuit may be shown in various ways, and a typical representation is given in Fig. 1-14(b). On occasion a square or oblong representation is utilized with the input and output leads identified by numbers which correspond to the number sequence shown in Fig. 1-14(a). All terminals may not always be utilized, and only those terminals actively connected to circuitry are shown in the schematic. If more than one IC is present, the schematic from the manufacturer usually includes a reference number for each IC such as IC1, IC2, etc. Often the IC is placed into appropriate holes in the printed-circuit board. The prongs are then bent over on the underside and soldered into place. Obviously an arrangement of this type presents some problems when replacements are necessary because the prongs must be unsoldered and bent straight so they can be lifted out. This procedure, plus the insertion and resoldering of the replacement, must be done carefully so that the printed-circuit conduction lines are not damaged. (See Sec. 10-4.)

1-8. VIDEO-AUDIO DETECTORS

The IF stages in VCRs and television receivers must accommodate not only the picture IF signal but also the sound IF signal. This is necessary because the tuner oscillator signal is mixed with both the video and sound signals, and hence two distinct IF signals are produced. As mentioned earlier, the standard television IF signal has a frequency of 45.75 MHz, and the sound IF signal has a frequency of 41.25 MHz. Prior to the detector sections these two IF signals are mixed in the last IF stage circuit to produce a 4.5-MHz *difference-frequency* signal. This 4.5-MHz signal now represents a new sound IF signal. At the output of the last IF amplifier stage the video and sound IF signals are applied to separate circuits for detection, as illustrated in Fig. 1-15. Here the video IF signal (45.75 MHz) is applied to the video detector, and the output now consists of the original video signal that was used to modulate the video carrier. In a television receiver the detected video signal is amplified and applied to the picture tube. In the VCR the video signal is applied to two sections, one consisting of the video output jack on the panel of the VCR and the other the recording processing section. The output 4.5-MHz sound IF signal is applied to a signal amplifier section and then to a frequency-modulation detector. In a television receiver the audio signal thus obtained is amplified additionally and applied to a loud-

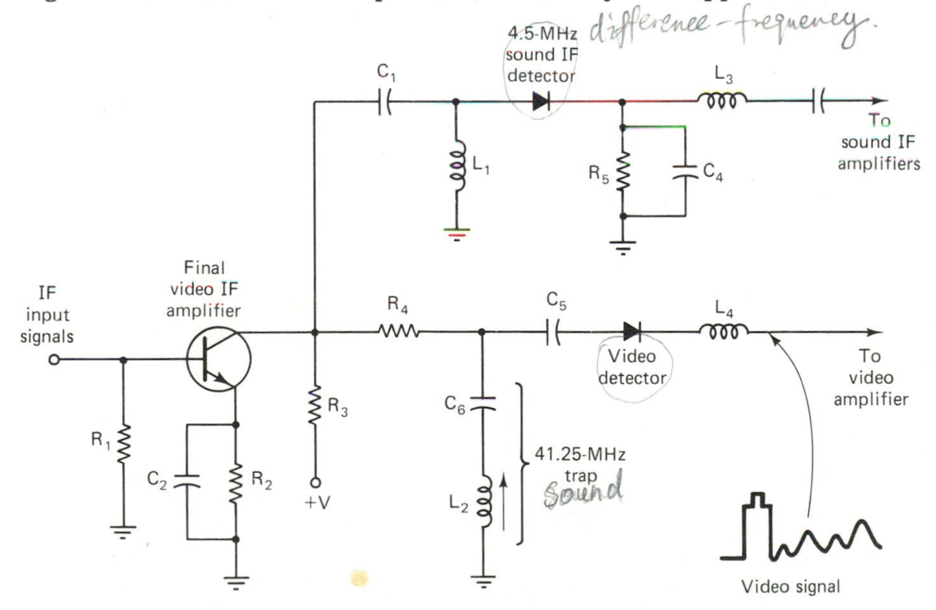

Figure 1-15 Picture and sound detectors.

speaker system. In the VCR the audio signal is applied to an output audio terminal on the rear (or front) panel (see Fig. 2-6) as well as to the appropriate VCR tape recording sections, as discussed more fully later.

Note that the video detector shown in Fig. 1-15 is a simple diode resembling the rectifier types. Since the signal amplitude handled by a detector diode is low, the diode is a small low-power type. The detector produces a series of single-polarity alternations that have an amplitude variation corresponding to the video signal. Capacitors and resistors convert this to a varying dc-type signal. When the latter is coupled to the next circuit (with a capacitor), the true video signal with ac characteristics is obtained.

Capacitor C_6 and L_2 form a trap circuit for the 41.25-MHz sound IF signal used with the 45.75-MHz picture IF signal to produce the final 4.5-MHz sound IF signal. The trap shunts to ground a sufficient amount of the 41.25-MHz signal to prevent interference in the picture tube. The arrow at L_2 indicates an adjustable variable-core coil for tuning purposes. The core has a screwdriver slot that permits turning to right or left to eliminate vertical-bar interference by the 41.25-MHz signal.

1-9. VCR MONITORS

Generally the VCR units are connected directly to television receivers with a switching-mode capability as detailed and listed in Sec. 1-3. When the VCR feeds a television receiver, the television signal travels through the tuner and IF sections of both the VCR and the television receiver. Consequently some deterioration of picture quality is encountered, though usually not to the degree where it would be objectional. If, however, a copy is made of a tape already recorded (see Fig. 2-6), picture quality may suffer to the degree where it will be undesirable. Picture quality can be improved by using a monitor instead of a television receiver. The monitor does not have a tuner or IF stages and accepts signals from the video and audio output terminals of the VCR. With a monitor the VCR must be used for tuning to different stations, and the advantage of recording one station while viewing another is lost with this type of monitor.

Since the quality of a video recording is judged by what appears on the screen of a television receiver or monitor, these two units may very well be contributing to poor picture quality if the latter is present. Because the receivers and monitors have adjustable levels of contrast, brilliancy, color, and tint, much depends on the proper setting of these controls. If careful adjustment of color control cannot correct picture faults, additional tests of the set must be made before assuming that the fault lies with the VCR. Since, however, television receivers and monitors are an integral part of the whole VCR package, one should be familiar with the general operations of the various sections to better evaluate the possible faults. It is not necessary that all the underlining technical and theoretical aspects be learned so long as one understands the purpose for a particular circuit and the problems that may arise when a particular circuit malfunctions.

The basic circuits of a television receiver and monitor are shown in Fig. 1-16. Those circuits which are not present in the monitor are identified by a black triangle at the lower right-hand section of the circuit block. Thus, it is evident that the monitor duplicates almost all the circuits of a television. Missing are the tuner, IF stages, sound and video

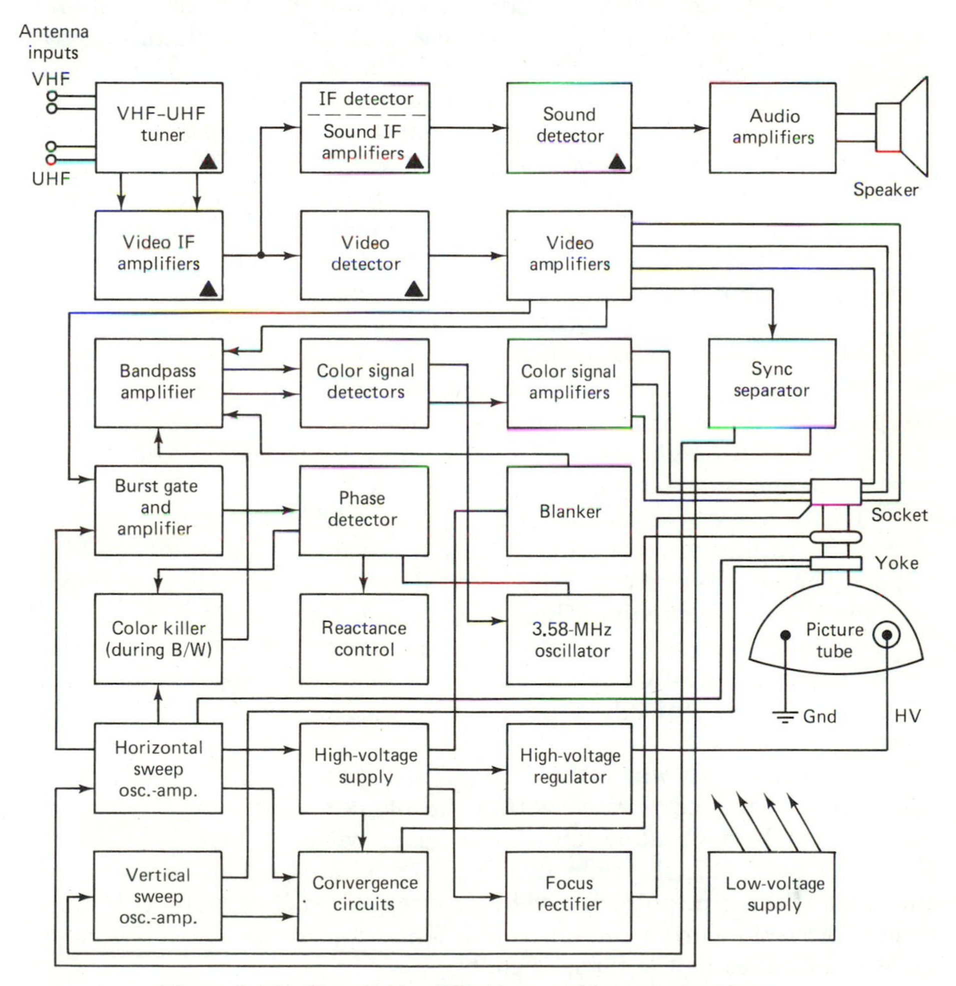

Figure 1-16 Circuit identifications and interconnections.

detectors, and sound IF amplifiers. With these circuits missing, the video and audio output terminals of the VCR are then fed directly to the audio amplifiers and video amplifiers. As shown later, contrast and brilliancy controls are included in the video amplifier sections.

The video amplifier section of a receiver or monitor feeds many other circuits, as shown in Fig. 1-16. The black-and-white luminance portion of a color signal is fed to the picture tube socket and then to the appropriate tube elements. The luminance, or brightness signal, is also designated as the *Y* signal. (See Appendix C.) These signals are mixed within the picture tube with the color signals obtained from the color signal amplifiers.

The video amplifier also feeds a circuit known as a sync separator. The latter strips the synchronizing pulses from the picture signal and transfers them to the horizontal and vertical sweep oscillators (signal generators). The video amplifier also feeds a bandpass amplifier and a burst gate circuit as shown. The bandpass amplifier processes the color signal sidebands. These sidebands are mixed with a 3.58-MHz carrier signal generated by a special oscillator described later. The reason for this is to reintroduce into the bandpass amplifier a replacement carrier signal for the one that was suppressed at the television station. This restores the complete amplitude-modulated video signal necessary for detection. A color killer circuit is used to make all the color circuits inoperative during the reception of a black-and-white signal. The killer thus minimizes interference during black-and-white reception. Color signal detectors follow the bandpass amplifier and form the three primary color signals of red, blue, and green (the additive color principle). These signals are then amplified additionally and also applied to the picture tube elements via the socket.

The necessity for creating a substitute 3.58-MHz carrier signal requires the formation of a group of circuits termed a *phase-locked loop* (PLL). The latter consists of the *phase detector*, the *reactance control*, and the *oscillator* circuits. The phase detector samples a portion of a transmitted signal having the precise frequency of the color carrier used at the broadcast station. This portion of a signal consists of eight cycles having a frequency of 3.58 MHz and is termed a *burst* signal because of its short duration.

The manner in which this burst signal is transmitted is somewhat innovative. It is incorporated within the blanking and sync signals that control the horizontal sweep of the beam within a picture tube. As shown in Fig. 1-17, the eight cycles of the burst signal are mounted on the right of the sync pulse on the horizontal blanking section. This burst signal forms the carrier reference frequency, but it does not have sufficient duration to be the carrier signal.

As shown in Fig. 1-16, the phase detector circuit samples both the

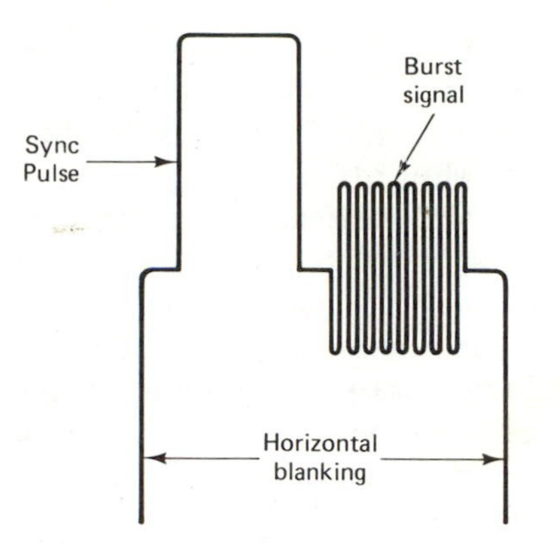

Figure 1-17 Burst signal.

burst signal and the signal generated by the oscillator. If both signals coincide in frequency and phase, nothing happens. If the 3.58-MHz oscillator drifts, however, the phase detector generates a correction signal which is applied to a reactance control circuit. The latter changes the tuning of the oscillator for a change of voltage input. Thus, the phase-locked loop maintains rigid frequency control. The oscillator is thus locked in to produce a substitute carrier identical in phase and frequency to the original carrier that was used but not transmitted.

The horizontal and vertical sweep oscillators generate the signals which sweep the beam horizontally, as shown earlier in Fig. 1-7, and also pull the beam down vertically for 525 lines per frame. In black-and-white receivers the horizontal scan frequency is 15,750 Hz, while for color it is 15,734.264 Hz. The vertical sweep frequency for black-and-white pictures is the same as the ac power lines (60 Hz). For color, however, the vertical sweep frequency is 59.94 Hz. There is such a slight difference between the black-and-white and the color sync frequencies that no trouble is encountered in maintaining good frequency control.

The horizontal sweep signal is amplified and transferred to the horizontal sweep coils of the yoke of the picture tube as shown in Fig. 1-16. Similarly, the vertical sweep signal is amplified and applied to the vertical coils of the yoke as also shown. The horizontal signal is also utilized to form the high voltage for accelerating the beam inside the picture tube. The voltage is stepped up and maintained at an acceptable regulated level. The high voltage is transferred to the inner conductive coating of a picture tube via an input terminal molded into the glass. There is also an outer conductive coating which is grounded as shown.

The glass between the inner and outer conductive coating of the picture tube acts as a dielectric to form a capacitor for filtering the ripple from the high voltage. It is this capacitive effect which stores the high voltage and can produce a severe shock even when the television or monitor is disconnected from the power mains. The picture tube retains this high-voltage charge for a considerable time after the power to the receiver has been shut off.

The horizontal and vertical scan signals are also applied to a convergence circuit that regulates the precise pinpointing of the three beams within a tri-gun color picture tube. A convergence control panel is usually present for making adjustments. Proper convergence adjustments produce the necessary overlapping of the three primary colors to produce white areas on the screen. Additional data on instruments used for adjustments are given in Chapter 7. Single- and in-line-gun picture tubes are also in use with preconvergence adjustments fixed and not requiring additional readjustments.

A tap on the high-voltage supply is utilized for the focus-rectifier circuit that creates an intermediate high voltage which is applied to the focus element within the picture tube. An independent control on the rear of the receiver permits adjustment of focus for maximum picture sharpness. Many receivers have a preset precise focus automatically maintained and requiring no manual adjustments. A low-voltage power supply system is utilized for activating all the various circuits illustrated in Fig. 1-16. The low-voltage power supply is similar to the types discussed in Chapter 8. Actual circuitry of sections illustrated in Fig. 1-16 are covered in later chapters where signal tracing and other servicing procedures are discussed.

The brilliancy control in a color receiver or monitor affects the luminous portion of a picture. Individual controls are present (usually available at a rear panel) for individual adjustment of the red, blue, and green beam intensities. A service switch is provided, as shown in Fig. 1-18, for checking individual beams. When the service switch is turned on, it eliminates the vertical scan, and a horizontal line appears on the screen. If proper convergence and correct beam intensities are present, a uniform white line should appear. If the red and green controls are reduced, a blue line should be visible. (The brilliancy controls should be reduced sufficiently so that the horizontal line is not excessively bright.) With the blue and green turned down, a red line should appear, and with the red and blue reduced, a green line should appear. If one color does not appear, turn up the gain control for that particular color. If the individual color cannot be produced, it may be caused by a defective picture tube. Failure of any section of the picture tube which kills a particular color causes incorrect color reproduction and the inability to produce white areas in a picture. Additional servicing data regarding color circuits will be found in Chapters 7, 9, 10, 11, and 12.

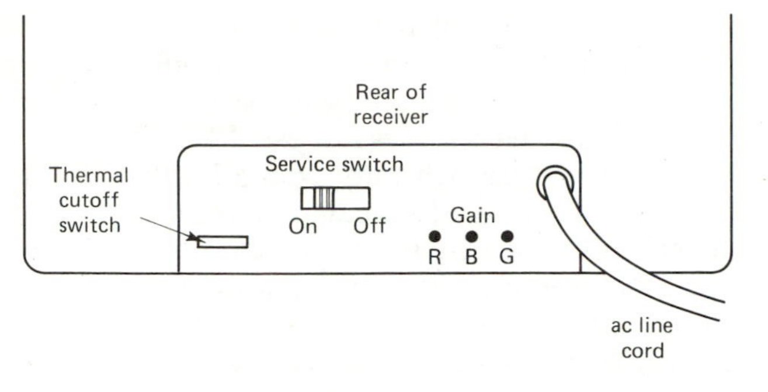

Figure 1-18 Service switch location.

1-10. QUICK CHECKING OF FAULTY PICTURE

Because the television receiver or monitor is the visual display accessory for the video tape recorder, it must be maintained in good operating condition to reproduce faithfully the sound and picture information recorded on tape. Generally, during playback, it is presumed that the TV operator controls such as contrast, brilliancy, color, and tint have been adjusted to provide a good picture. If the television receiver produces good pictures and sound during reception of antenna or cable stations, it should also perform satisfactorily when playing a video prerecorded tape.

If the picture quality is poor during tape play or when viewing a picture tuned in by the VCR, it is not necessarily a component or circuit defect in the VCR or television. Hence, several quick checks should be made to eliminate normal possibilities before making extensive chassis and component tests and measurements. Often the fault may be a loose connection, defective cable linkage, or similar problems associated with other external accessories. Once all logical possibilities have been exhausted, the need for testing and troubleshooting circuit and tape transport mechanisms has been validated. If the quick-check procedures have localized the problem, much needless and time-consuming test procedures will have been avoided.

Assume, for instance, a video tape recording has been made, but on playback the picture displays interference streaks across the screen. In such an instance any sequence of several quick checks can be made. If another recording is on the tape, it can be played for comparison purposes. If the streaks are not present, you have pinpointed the problem as existing only in the segment recorded last. The possibility then exists that the interference may have been present in the signals received by the antenna or cable. If the picture had been viewed on the receiver while recording, any streaks during reception would have been noticed. If the tape has been repeatedly used, there is a possibility that the tape section may be worn and hence does not record properly.

Another possibility, though rare, would be that the television receiver developed a fault at the particular time the recording was made. This can be checked by viewing several stations or recording something and replaying it. Cable connections should also be checked (see Chapter 2) since a loose connection could cause interfering streaks. The problem could also be caused by recording heads that need demagnetizing and cleaning. (See Chapter 3.)

If the picture is weak for tape playback or direct viewing but without streaks, the contrast and brilliance controls may need readjustment on the receiver. The incoming antenna or cable connection can also be removed from the VCR and applied directly to the antenna input terminals of the television receiver. If the picture quality is still poor or the signal is weak, the antenna or cable input system must be checked. If another receiver is available, a quick check can be made by attaching the input lines to it and checking several stations. If the picture quality is now good, the fault would be in the receiver. If the quality is still poor, the fault lies with the antenna or cable input system.

Faint lines or images from another station during playback of recorded video tapes are often caused by adjacent-channel interference from either an upper or lower channel (see Fig. 1-10). Television station frequencies have been allocated to avoid such adjacent-channel interference in any particular vicinity. Thus, if a station for channel 3 is in the area, channel 2 or channel 4 would be in another area at a sufficient distance to minimize interference. During abnormal atmospheric conditions, however, an adjacent channel from another area may be sufficiently strong in your area to cause some interference. When this is the case, such interference will, of course, be recorded along with the desired information.

Adjacent-channel interference can, of course, also occur for other channels, though not all are adjacent in terms of frequency even though they are adjacent numerically. (See Appendix A.) Channels 6 and 7, for instance, are not adjacent in frequency since the FM radio band is sandwiched between them. Similarly, there is a frequency gap between channels 4 and 5 as well as 13 and 14.

The separation between channels 3 and 4 is of particular significance because the VCR output terminal that is linked to the antenna input terminal of the TV is controlled by a switch to select an output signal for either channel 3 or 4. This means that the television receiver must be set on 3 or 4, *whichever has no broadcast in the area*. The VCR switch is then set to coincide with the chosen station. The same factor is utilized in connecting a home computer to a television. Like the VCR, the home computers have an output for either channel 3 or 4, and the output is selected for the one that is not broadcast locally.

A similar condition can occur with cable television. A cable company also should not transmit on both channels 3 and 4 to avoid the interference which could affect VCR recording and playback. Even if the recording has no interference in it, interference prevails, it can be remedied by persuading the cable company to leave either channel 3 or 4 open or by purchasing a selector switch box available from jobbers. As shown in Fig. 1-19, such a box will permit you to disengage the incoming cable during playback and thus avoid the interference which can mar a good recording. Even if one terminal is unconnected as shown, the antenna input is switched away from the VCR during tape playback.

A switch such as shown in Fig. 1-19(a) has a 75-Ω impedance at each terminal. The switch can also be used in reverse, as shown in Fig. 1-19(b), if necessary. Thus, an outdoor antenna and a commercial cable line can be applied to the dual terminals and the single terminal used to feed a VCR. This arrangement permits use of an antenna during the time the cable TV may fail or when it is temporarily under repair. More elaborate switching boxes are available with additional input and output terminals as needed. Some jobbers also carry amplifiers to boost weak input signals. The switches are generally designed to avoid interaction between the switch terminals by designing them with adequate spacing and more than usual switch lever travel. Although ruggedly de-

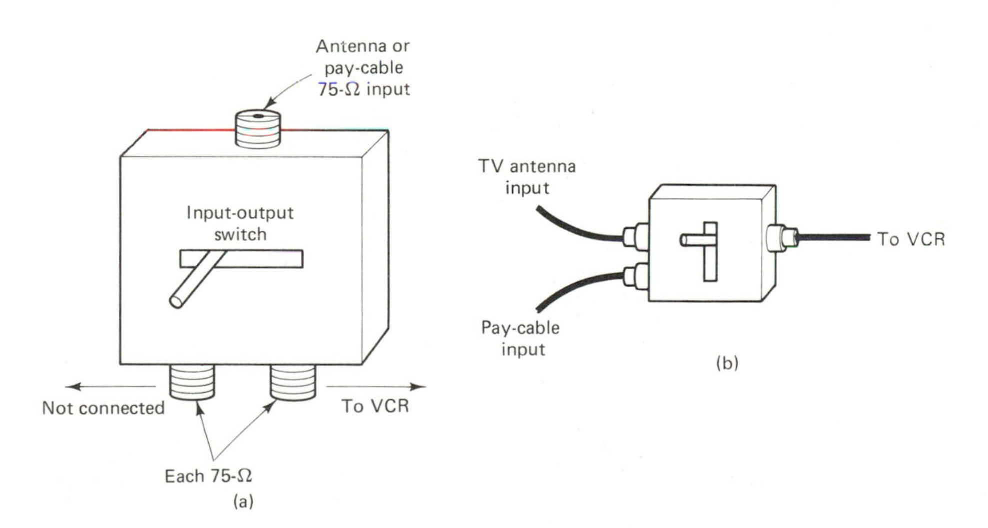

Figure 1-19 TV switch applications

signed, some faults could develop with the switch assembly, and this factor should be considered if problems suddenly occur when switching from one section to the other. Make sure all connections to and from the switch are tight and that the coaxial lines are in good working order.

2

Problems in VCR/TV Linkages

2-1. VCR/TV INTERCONNECTIONS

A VCR must have an input terminal for connections from the antenna (or the cable) system, as shown in Fig. 2-1(a). An output terminal is provided for connecting the VCR to the antenna input terminal of the television receiver. For versatility, however, the VCR must be capable of switching the incoming signal directly to the TV receiver without recording, or it must be capable of switching the incoming signal to the tape recorder mechanism as required. The variations in the switching modes have been detailed earlier in Sec. 1-3. Usually the provisions are for coaxial cable input. If the antenna lead-in consists of a twin lead, an adapter such as shown in Fig. 2-1(b) must be utilized to convert the 300-Ω twin lead to the 75-Ω coaxial cable output line. Failure to do this results in an impedance mismatch and inferior results.

If the TV antenna system has a dual twin-lead arrangement for UHF and VHF, an adapter such as shown in Fig. 2-1(c) must be used for converging both leads to a single coaxial cable input to the VCR. If the antenna line is a coaxial cable (or a paid-cable-line input system is utilized), no matching sections need be used. The connections from the output of the VCR to the input of the television receiver should also be a coaxial cable linkage. If the television receiver has a twin-lead input terminal, the adapter shown in Fig. 2-1(b) can be used in reverse, with the coaxial line from the output of the VCR connected to the coaxial input of the adapter. The twin-lead section would be connected to the television input.

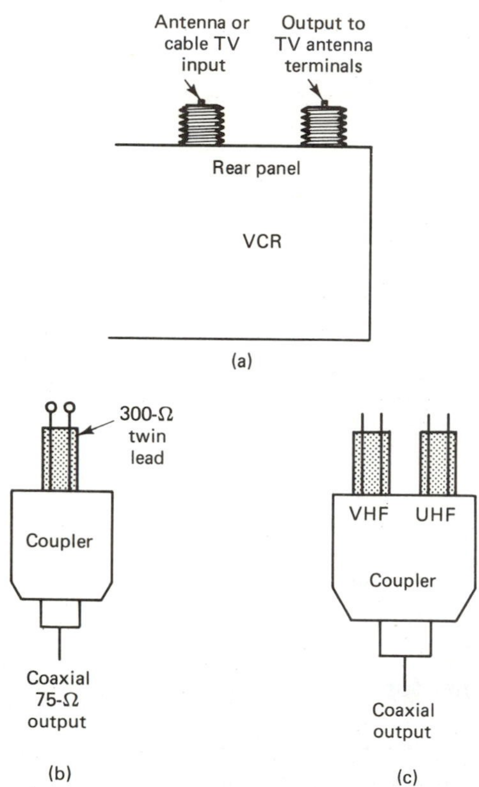

Figure 2-1 Connections and couplers.

2-2. COAXIAL CABLE FACTORS

The basic structures of coaxial cable plugs and sockets are illustrated in Fig. 2-2. Coaxial cables are extensively used in all linkages of video cassette recorders to antennas and television receivers as well as in microphone cables and other electronic devices. Such cables often develop faults, particularly if they undergo constant flexing as equipment is moved about. In particular, the most trouble-prone cables are those associated with microphones and cable linkages between portable VCR units and movie cameras. Replacement cables complete with plugs are widely available from jobbers and radio stores. Some of the latter also stock cable reels and sell required lengths as well as appropriate plugs and sockets.

Most coaxial cables are made up of the sections shown in Fig. 2-2(a). Here an inner copper wire forms one of the two conductors essential to transfer information in electric form from one point to another. The other conductor is the outer metal braid covering of the

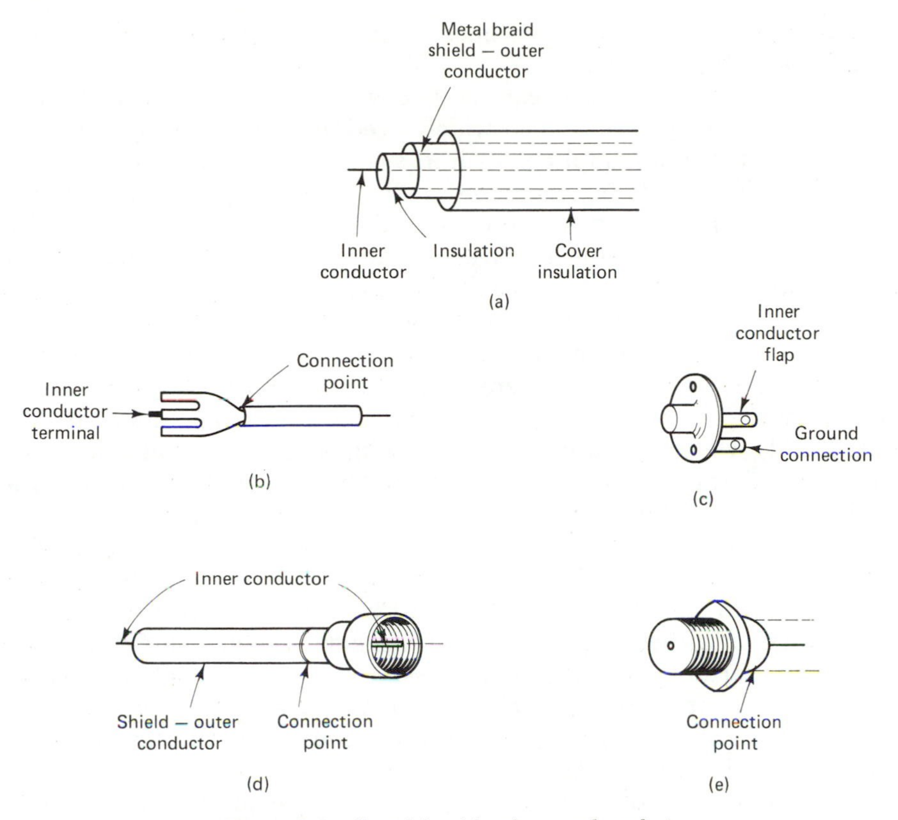

Figure 2-2 Coaxial cable plugs and sockets.

coaxial cable. A low-loss dielectric plastic material is used between the inner conductor and the outer metal braid conductor for insulation purposes between the two. An additional plastic outer cover is used for insulation of the outer conductor ground terminal from other devices. The impedance of the most commonly used coaxial cables varies between 50 and 75 Ω. Most of the VCR and television coaxial cables have a 75-Ω impedance as compared to the 300-Ω impedance for the common twin-lead conductor often used between the antenna and the television receiver.

For the VCR and TV purposes the coaxial cable is preferred over the twin lead because it has a minimum noise pickup characteristic. The twin lead has less losses per foot of length as compared to the coaxial cable, but this advantage is overcome when comparing the twin lead to the low-noise and interference elimination characteristics of the coaxial cable. Since the two transmission lines have different impedances, they should not be connected directly to each other. The mismatch between the twin lead and coaxial cable would introduce losses. Instead, when it is necessary to couple the two lines together, the units shown earlier in Fig. 2-1 should be used.

One type of plug that can be used with coaxial cables is shown in Fig. 2-2(b). This plug has been widely used in phono and tape recorder linkages to the main amplifier of hi-fi systems but is also used to some extent in VCR units for video and audio input and output lines. (See also Sec. 2-4.) The common socket used is shown in Fig. 2-2(c). This plug and socket make for a convenient and quick connect and disconnect process. This feature, however, also has the drawback that the plug can work loose from the socket and impair reception.

The connection point shown in Fig. 2-2(b) and (c) indicates where the coaxial cable connections are made to plugs and sockets, as detailed in Sec. 2-3. The standard coaxial cable screw-on plug is illustrated in Fig. 2-2(d). Here the outer shell of the plug is threaded and a stiff-wire inner conductor section is present. When this plug is threaded onto the socket shown in Fig. 2-2(d), a tight connection is formed that resists being disengaged when units are moved and cause cable flexing or shifting. The connection points indicated in Fig. 2-2(d) and (e) are the places where the outer braid is stripped back and crimped or soldered around the outer flange of the plug and socket, as detailed in Section 2-3.

With portable VCR units and associated devices such as cameras and stereo units, some special cables may be encountered. Some of these cables and plugs have provisions for *dual input* (DIN), though the plug and socket units may have provisions for several inner-conductor lines, using a single outside metal braid as a shield and common ground conductor. Typical is the unit shown in Fig. 2-3. These units are made with three-prong or five-prong plugs and matching sockets, and on some occasions more than five inner conductors may be present. The design of the plug and socket is usually such that the three-prong plug will also fit into the five-prong socket. With a cable containing several inner-conductor wires (each insulated from the other), any that are suspected of being defective require a continuity check of each individual inner conductor. (See also Chapter 7, Fig. 7-8).

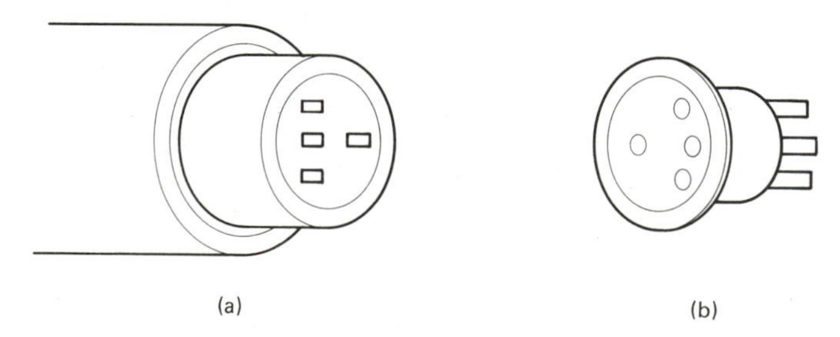

Figure 2-3 DIN plug and socket.

2-3. CABLE REPAIR AND REPLACEMENT

Coaxial cables complete with end plugs are available from jobbers and electronic supply stores. Lengths are available from short sections to a roll of several dozen feet. Cables that develop problems are those that undergo some flexing such as by moving the VCR or shifting linkage positions. Cables used with portable VCR units are particularly prone to developing problems. It is advisable to have a 6- or 12-ft length of spare cable on hand, complete with end plugs. Such a spare cable can be used as a temporary substitute when making the quick checks detailed in Sec. 2-4. The spare cable can also serve as a temporary replacement while another of proper length is being prepared for permanent linkage. Although short lengths of coaxial cable with plugs are readily available, specific lengths can be prepared as desired by obtaining cable and separate plug fittings and attaching the plugs to each end. It is also advisable to become familiar with the procedure for plug replacement so that repairs are expedited when some defect occurs at the point where the cable is attached to the plug.

In preparing the end of a coaxial cable for plug attachment, the outer plastic covering must be removed for a short distance from the end of the cable, as shown in Fig. 2-4(a). After the outer covering has been removed from the end, the metal braided shielding must be cut back to expose the inner-core plastic insulation. As shown in Fig. 2-4(b), the end of the plastic insulation is also cut back to expose the inner-conductor solid wire.

(Some microphone cables will use stranded wire to minimize breakage because of the constant flexing. The plugs and sockets are similar except that the inner conductor in the plug ends where the plug threads start. The soldered conductor end makes contact with a similar soldered-wire lump in the socket. For VCR and TV linkages, the standard solid-wire cables should be used.)

The inner plastic section should be stripped from the wire only to the point where the inner solid wire goes through the insulated hole in the plug and appears in the outer ring, as shown in Fig. 2-4(c). There should be sufficient inner plastic insulation between the copper wire and the braid to prevent a short circuit. Thus, make sure *not a single strand of the braid wire* touches the bare copper wire. Once the plug has been inserted over the prepared cable, it might be a good idea to make a continuity check between the inner and outer conductor to verify that no short circuit exists.

If connections appear to be all right, the section containing the rolled-back braid is crimped with a pair of pliers or a crimping tool. In lieu of crimping, the section can be soldered, though the crimping is a

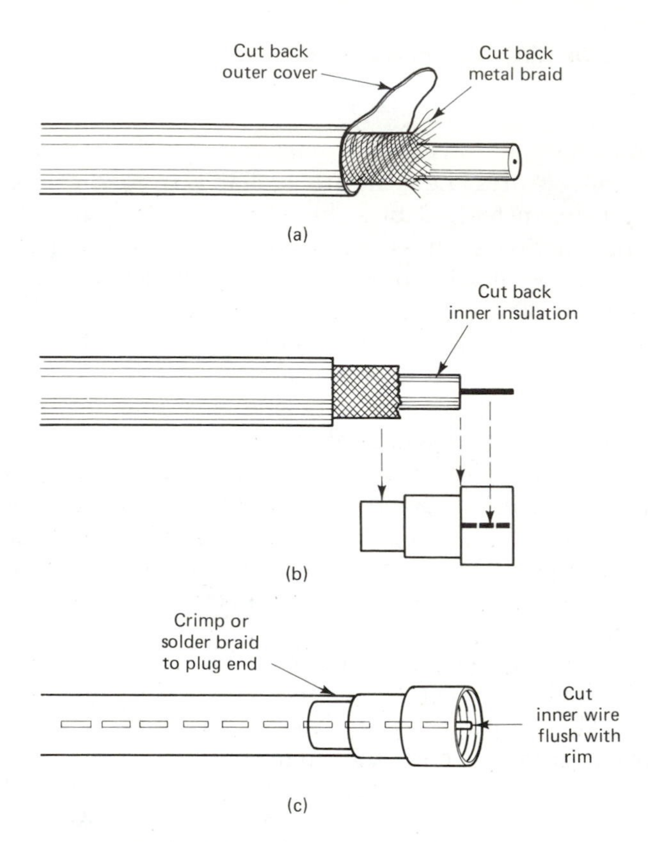

Figure 2-4 Attaching new plug units.

simple and effective procedure if it forms a firm grip between the end of the plug and the metal braid that was rolled back.

Usually the same type of plug is attached to the other end of the cable, and after the cable is finished, another continuity check should be made, as illustrated later in Fig. 7-8 (between inner and outer conductors and between each end of the inner conductor). Similar procedures are used to attach a socket to one end if this should be necessary, though in such an instance the coaxial cable is attached to the terminals available, and no wire protrudes through the front. [See Fig. 2-2(e).]

2-4. QUICK CHECKING OF CABLE LINKAGES

When poor reception (or lack of reception) occurs on the VCR and television receiver, the problem may be in the coaxial cable linkages to and from the units involved. With any type of reception problems or playback difficulties, the cable linkages should be checked initially to eliminate these as possible trouble sources. Once the cables have been checked out, more intensive checks can then be undertaken on the VCR and television receiver, as detailed in subsequent chapters. (see also Table 11-1.)

Two major symptoms of cable defects consist of intermittent performance from the VCR/TV or total lack of reception. With the intermittent type a quick check can be made by flexing (one at a time) each cable linkage slowly to see what effect there is on the picture reproduction (or recording). For instance, with a picture being received from the antenna (via the VCR) as shown in Fig. 2-5(a), grasp the cable near its

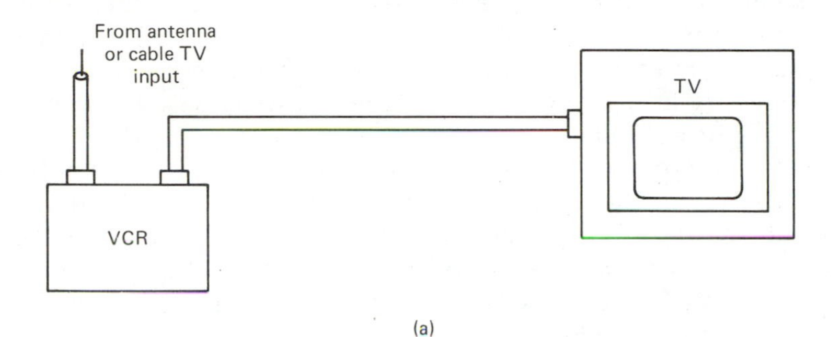

(a)

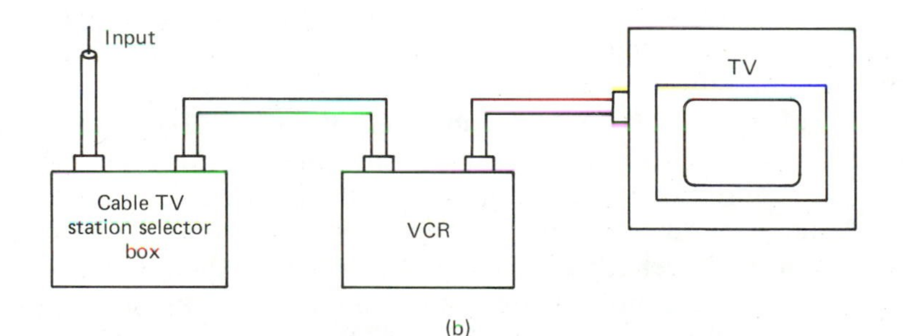

(b)

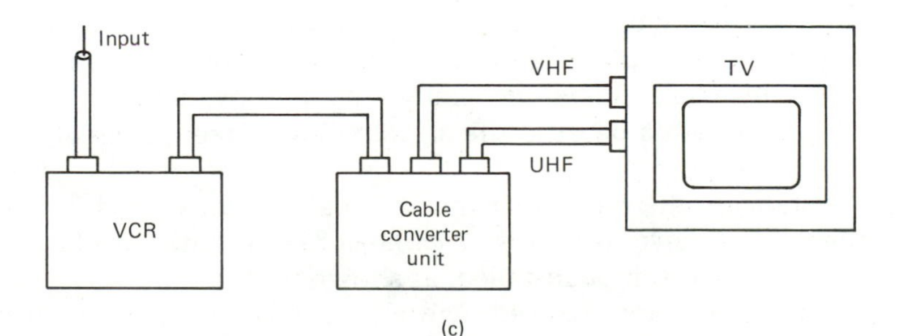

(c)

Figure 2-5 Various VCR to TV linkages.

entrance to the VCR and slowly flex it while watching the screen. An increase in the intermittent condition usually indicates the localization of the fault.

If the flexing at the antenna input terminal does not aggravate the condition, flex the connecting cable that runs from the VCR output to the antenna input terminals of the television receiver. Again note the effect, if any, that may appear on the screen. If these checks have disclosed intermittent conditions, the defective cable can then be repaired or replaced. Often the trouble may occur in a poor connection at the section where the cable braid is clamped to the plug. A corroded or oxidized braid wire may cause some problems, and this area should be checked. Similarly, the plug end with the inner wire extension should be examined carefully for broken or bent wire or poor connections to the socket. If a new cable is tried and the trouble still prevails, the socket may be at fault.

In commercial cable television a station selector box may be used, as shown in Fig. 2-5(b). The placement of this box in relation to the VCR and the television receiver depends on the cable station capability of the particular VCR or television receiver. If the reception capabilities of both the VCR and the television receiver are limited to 90 stations, some transmitted cable stations will not be available without using either the cable station selector box or a cable converter, as shown in Fig. 2-5(c). (See also Sec. 12-4.) With a cable selector as shown in Fig. 2-5(b), there are added coaxial cable links interconnecting the three units. In some installations the cable selector box may have coaxial linkages sufficiently long to permit remote control operation from across the room. In such an instance the linkages may be subject to more flexing and moving than normal and hence can create linkage problems. Again the quick-checking procedures consist of slowly flexing each cable along its length and noting the effect on the picture.

If the picture is intermittent and the cable checks out all right to and from the cable selector box, the problem may be in the linkage between the VCR and the television receiver. If the couplers illustrated earlier in Fig. 2-1 are utilized, the additional interconnections can cause some problems if the plugs are not tight. If the trouble persists, a new coupler can be tried, though generally very few problems are encountered with these units because the signal voltages are extremely low in amplitude.

In instances where the VCR has full cable television station capabilities but the receiver does not, a cable converter unit (available from radio parts jobbers) can be installed, as shown in Fig. 2-5(c). Here again, extra cable linkages are involved, and all must be checked out if intermittent conditions are present. If intermittent conditions prevail only for a particular station, however, the fault may be in that particular

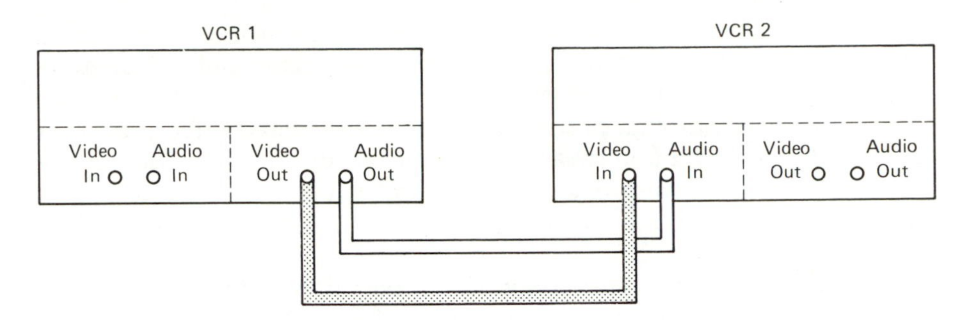

Figure 2-6 Linkages during copying.

station's transmission. With the mechanical-type tuners in older receivers, however, it is possible for noisy tuner switches to cause interference in one station but not in others.

Often, when a camera and VCR are used to capture family picnics, parties, weddings, and so on, requests are received for copies of the tapes. Thus, the setup shown in Fig. 2-6 is used, with VCR 1 containing the recorded tape and VCR 2 the blank tape for copying. When troubles occur in recording under these circumstances, the initial tests and checks should again involve the interconnecting cables. Since the hookup shown in Fig. 2-6 is only used for special purposes, a defective cable may not be suspected. (See also Sec. 12-3.)

If only the sound suffers during the recording, the two cable linkages can be interchanged and a sample copy recording performed. If, now, the video reproduction is intermittent or of poor quality, it verifies the fact that one of the cables is defective because it introduces trouble symptoms in both video and audio when interchanged. Since these cables have the type of plugs and sockets illustrated earlier in Fig. 2-2(b) and (c), a loose-fitting plug or socket may be the cause of the trouble. Make sure the four plugs are pressed firmly into the sockets when making the cable-flexing tests.

2-5. STEREO LINKAGE FACTORS

Early in 1984 the FCC approved a system for broadcasting stereo sound to accompany the video. As with stereo FM broadcast by FM stations the system is compatible. Thus, a mono broadcast received by a television equipped for stereo will produce a mono sound output. Similarly, a stereo sound broadcast with television received by a mono sound TV receiver will produce a mono output. Units are available, however, that will convert a TV receiver with mono sound into one capable of receiving and processing the stereo sound. In the latter instance true stereo reception is achieved, even with older receivers. This method of stereo reception must not be confused with the addition of a stereo synthesizer. The latter, though producing a simulated stereo effect, does not provide the exact sound separation as picked up in a television studio and broadcast. (See also Sec. 12-2.) For a VCR not designed to decode the broadcast TV stereo sound, an adapter is also available for conversion purposes.

The multiplexed stereo sound accompanying the video is also present in the normal cable linkage from the VCR to the input of the TV set's tuner, the latter set for either channel 3 or 4. When, however, the audio output from the stereo jacks is utilized (such as in the copying of tapes), additional problems in cable linkage may be present. Similarly, the dual-output audio stereo lines from a stereo-capable TV receiver increase the possibility of cable linkage problems between units. This is obvious when one considers the mono process: The VCR feeds the television during playback, and the single mono speaker of a television is utilized. With a good stereo system the television receiver must either have good high-fidelity stereo sound capabilities, with dual speakers, or the output audio lines must be connected to an external high-fidelity stereo system.

When troubles occur in stereo systems, quick-check procedures are useful and on many occasions disclose the defective sections or problems quickly. If, for instance, a VCR were linked to a television receiver, with both possessing stereo capabilities, there would be two output speaker systems involved, as shown in Fig. 2-7(a). Here, assume that the left-channel output is designated as 1 and that the right-channel output is designated as 2. If channel 1 is inoperative, for instance, a quick check consists of unplugging both lines 1 and 2 from the stereo output terminals of the television and then replacing them by plugging each into the opposite output jack. Thus if 1 now produces sound when plugged into the 2 socket and the line and speaker 2 is now inoperative, it immediately indicates that the connecting cables for each speaker are in good working order and also that the speakers are not defective. It would indicate, therefore, that the fault lies in the output circuitry or sockets that originally fed speaker system 1.

If, during the interchange of the dual cables and speakers, there is no sound from speaker 1 now in the opposite socket, it would indicate that the connecting line or speaker 1 is defective. Thus, a continuity check should be taken between the ends of the interconnecting cable to check for any open or shorted conditions, as previously discussed for linkages in earlier portions of this chapter. Similarly, if the cable is found to be in good working order, a continuity check should be made of the speaker to ascertain whether or not an open circuit exists be-

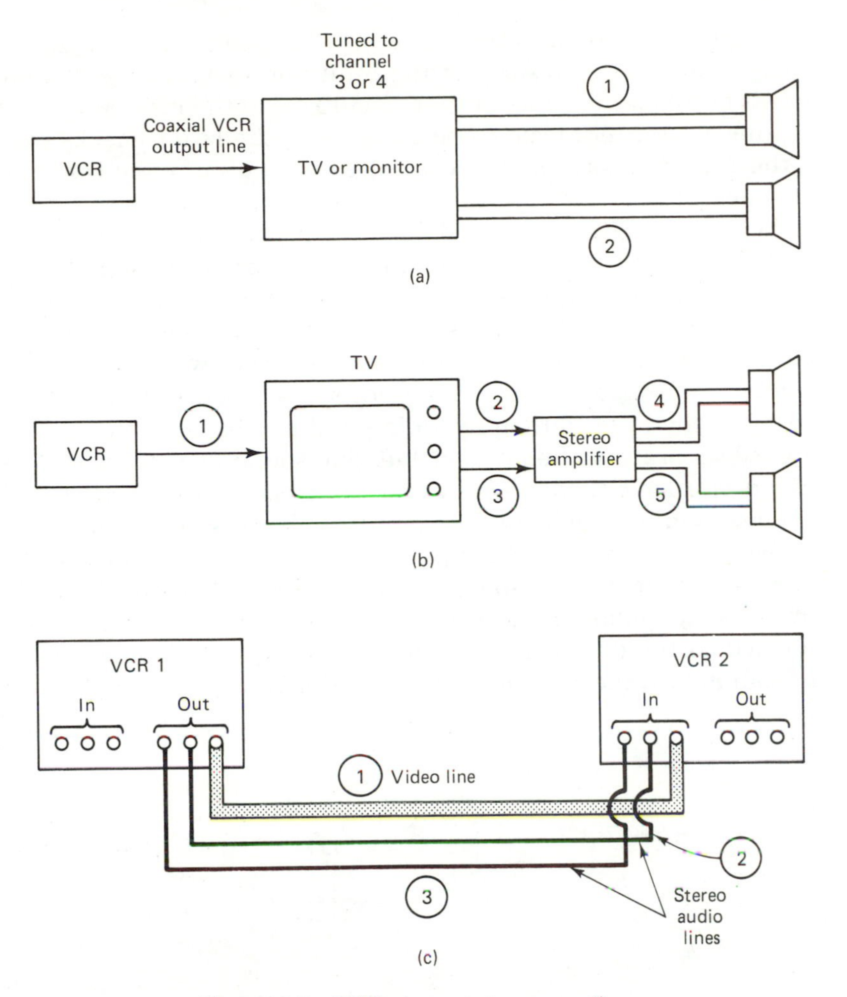

Figure 2-7 VCR stereo interconnections.

tween the speaker terminals and the internal speaker voice coil. A substitute speaker can be tried to verify the defect in the original speaker.

If the stereo TV feeds a separate stereo high-fidelity amplifier as shown in Fig. 2-7(b), the additional interconnections also must be checked. Similar procedures are undertaken as for the case shown in Fig. 2-7(a). For the system illustrated in Fig. 2-7(b) linkages 4 and 5 can be interchanged if one channel is inoperative. If the speakers and connecting wires are found to be all right but no sound is obtained at the output to speaker section 4, it is possible that the interconnecting cable 2 between the TV and the stereo amplifier is defective. Now cables 2 and 3 can be interchanged, if the speaker connected to line 4 now produces sound but the other speaker does not, the fault will be with the original interconnecting cable 2. If the cable linkage 2 does not show any defects when tested for continuity or open circuits, the problem exists between the audio output circuitry and the linkage inside the TV to the output terminals feeding line 2. If the latter checks out all right, the fault will have been localized to the output circuitry feeding this channel sound.

When two VCRs are utilized for tape-copying purposes as shown in Fig. 2-7(c), an extra cable again is involved for producing the stereo sound. Thus, if the playback of the copied tape discloses a weak or poor sound channel of the two stereo channels, the same quick-check procedures can be performed of interchanging the two audio cables to see whether or not lack of output or poor sound quality from one channel shifts to the other during recording and playback. If the interconnecting cables prove to be all right but one sound channel is still dead or has inferior sound, additional checks can be made before removing the cabinet and working on the circuitry. If possible, another VCR should be obtained and used for playing the tape that has been copied. If this results in poor or no sound from one channel, it would be highly probable that the fault lies in the second VCR. Thus, the third VCR can be attached to the output line and used for copying purposes to verify the indication that the recording VCR is defective.

3

Routine Maintenance Procedures

3-1. NECESSITY FOR LUBRICATION AND HEAD CLEANING

Maintenance procedures are essential in any system involving mechanical devices and linkages. All moving parts must remain lubricated so they do not bind and cause sluggish operation. These factors are particularly important in VCR units because of the mobility required in the video recording heads. Similarly, there must be free movement in the rachet assemblies that initiate the turning of the two tape handling spools in the cassette.

As with audio cassette recorders, the sliding tape across recording and playback heads eventually causes a slight accumulation of residue from the magnetic material impregnated on the tape. Such an accumulation causes a deterioration in the reproduction of the recorded video and sound signals. Not only is the playback function impaired, but the recording procedure is not as effective as it normally would be. Another consequence of tape sliding across recording and playback heads is the ultimate accumulation of magnetism on the heads. The latter can also alter the recording and reproduction of video and sound. The demagnetization factors are covered more fully in Sec. 3-5.

3-2. CLEANING METHODS

For proper head cleaning, it is advisable to obtain a cleaning fluid available at jobbers and parts distributors. Such a cleaning fluid has been formulated to minimize the possible damage that may occur to heads if strong chemical cleaners are used. Cotton-tipped swabs, so popular in audio cassette head cleaning, should not be used on video heads. Cotton strands can become dislodged and become wedged in the spinning video head ensemble. Loose strands can also damage some head mounts and cause other problems. For safe cleaning procedures, use a swab tipped in chamois or cellular foam. If jobbers can't supply these, use a swab made of any lintless cloth such as muslin. A small piece of chamois can be cut and used as a cleaning swab [See Fig. 3-1(a).]

Because a VCR contains a minimum of two video heads as well as control, erase, and audio heads, cleaning procedures are more involved than is the case with audio cassette tape head cleaning. As discussed in Chapters 5 and 6 the two video heads are mounted on opposite sides of a spinning cylinder or drum. For a good head-cleaning procedure it is essential that the outer cabinet of the VCR be removed, taking all precautions detailed in Sec. 3-3.

When four or five video heads are in use, the cleaning procedure is somewhat more involved, as detailed later. Although excellent results can be achieved by removing the cabinet for direct access to the heads, an alternate method which eliminates the necessity for access into the VCR is to use a head-cleaning cassette available from jobbers. Such a tape-cleaning cassette has a chemical on the tape surface, and when the cassette is played in the VCR, the heads are cleaned as the tape slides over them. Two types have been generally available; one is the dry chemical type, and the other uses a moist compound.

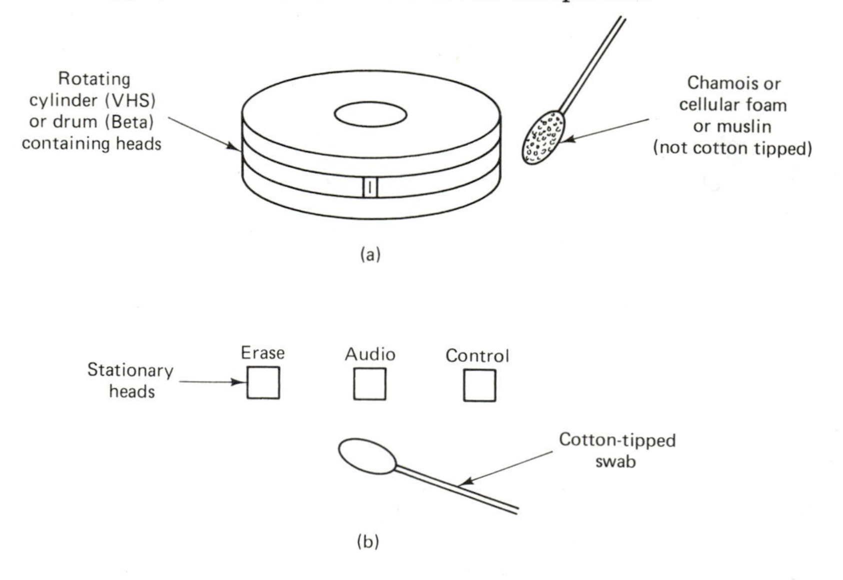

Figure 3-1 Cleaning swab types to use.

The use of swabs and chemical cleaners directly on the heads is preferred in those instances where the general dusting of the internal parts is to be performed or when bearings are in need of periodic oiling. If the metal cabinet of the VCR is to be removed, the factors outlined in Sec. 3-3 should be observed. These precautionary measures are of extreme importance in minimizing shock hazard. Eject any cassette to unwrap the tape from the drum for cleaning access.

In moistening the swab for head cleaning, use only commercial cleaners designed for this purpose. Employing other chemical cleaners on the recording heads may cause damage. Do not use tweezers or other metal devices to hold the cleaning swabs because of the danger of scratching the heads. Avoid touching the polished drum or cylinder that contains the heads.

The swab should be moistened with the cleaning fluid but not saturated to the point where it could drip down from the head into the VCR compartment. Recheck operation after the heads have been cleaned, and if poor quality reproduction appears to be linked to the tape head process, check for worn tape (discussed more fully in Sec. 3-4) and also consider demagnetization, as discussed in Sec. 3-5.

With more than two heads on a drum or cylinder, moisten the swab and use a sideway cleaning motion for a head area. Move the drum by turning it with a lint-free swab and clean a new area. Continue the process until all head areas have been cleaned.

Cotton-tipped swabs can be used to clean the stationary audio erase and control heads, as shown in Fig. 3-1(b). With a cassette equipped with a cleaning tape all heads are cleaned simultaneously.

3-3. CHASSIS ACCESS PRECAUTIONS

Of extreme importance when removing a VCR housing is to observe all necessary precautionary measures to avoid electrical shock. Since oiling, cleaning, demagnetizing, and greasing procedures do not require the VCR to be in operational readiness, the line cord should be removed from the wall socket. Do not rely on the on-off switch; that is, don't assume that since the switch is in the off position the line cord can remain plugged in. When the chassis of the VCR is exposed, there are several sections that are linked to the line cord entry point, and these could still produce a severe electrical shock. Even when the line cord has been removed, the electrolytic capacitors in the power supply system may still be holding a charge that could produce a severe shock.

In removing the housing, first take out all holding bolts or screws. Once the cabinet no longer is held, remove it carefully. Often pushbutton sections can hamper easy removal, and the housing must be shifted to permit clearance of all external access controls. After cabinet removal, note the location and position of the heads. Do not try to rotate the video cylinder or drum by hand if all heads are not readily accessible. Body chemicals on the skin can corrode the polished ring that contains the recording-playback heads.

3-4. TAPE QUALITY

As a video cassette tape is played repeatedly, tape wear occurs, and the smooth coating of magnetic material will wear off in spots. Consequently, the reproduction suffers, and the picture quality gradually deteriorates in sharpness. A sufficient wear on the tape will also cause interference streaks to appear on the screen. A tape of this type should not be played because it causes excessive wear on the heads. The wear is caused because the tape's surface is no longer as smooth as it should be. Not only does such worn tape affect tape head wear, but it also results in a greater accumulation of magnetic material residue on the tape head. Hence, the heads require cleaning more often. If the recorded material on any tape is such that it is to be retained, it should be recopied when the first signs of deterioration occur.

Video cassette recorders permit selection of a slower recording-playback speed than normal to obtain longer-duration programs. The Beta system uses Beta II as the normal speed for best reproduction and Beta III, the slower speed, for extended play. The Beta I speed is no longer used in home VCRs. The VHS machines have three speeds: the SP (standard play) is the fastest, the LP (long play) is the middle speed, and the SLP (super long play), also known as EP (extended play) is the slowest. With either the Beta or VHS units, use of slower speed introduces some impairment of picture quality. Depending on the type of program, the decreased picture quality may not be too apparent but is present nevertheless. As the recorded tapes are replayed numerous times, the initial lower-quality recording will manifest greater deterioration as wear accelerates.

Blank tapes also are available in longer-play tapes. The Beta L-750, for instance, holds 3 hr of recording in the B II speed setting and 4 hr and 30 min for the B III speed. The L-830 tapes give 3 hr and 20 min at B II and 5 hr for B III. In the VHS system 8 hr of recording can be obtained on a single tape. The VHS tapes are designated as T 120, T 160, etc.

In purchasing blank VCR tapes, it must be realized that the longer play reels are composed of thinner tape and that greater wear might be experienced in comparison to thicker tape. Even for the same length tapes, however, there is a difference between the tapes of one manufacturer and another. Some tapes may deteriorate after 30 or 40 playbacks, while others may last for 70 playbacks. When a VCR is first put into operation, it may be advisable to vary the brands of tapes that are purchased initially and finally use only those that exhibit longer life and better quality for a particular tape length reel.

3-5. HEAD DEMAGNETIZATION

In all cassette recorders, whether audio or video, the recording and playback heads as well as the erase heads ultimately require demagnetization. During playback, the heads slide across the magnetized sectors of the tape and eventually acquire some magnetism (residual magnetism). The latter condition requires demagnetization because a magnetized head will interfere with good recording and playback. The ability of the heads to record or play back higher-frequency video and audio signals may become sufficiently impaired to be noticeable in both the picture and sound during playback. When this occurs, the head demagnetization (also termed *degaussing*) should be undertaken.

The heads are demagnetized by using a portable electromagnet available at jobbers. The unit is plugged into the ac line socket and turned on with a push button. The unit has the general appearance shown in Fig. 3-2. Access to the heads requires removal of the metal cabinet (outer enclosure) of the VCR to expose the recording heads. Not only should the rotating video record/playback heads be demagnetized but also the erase head, the audio head, and the control head. (See Fig. 3-1.) Since the demagnetization process does not require the VCR to be turned on, the VCR line cord should be removed from the socket as a precautionary measure against shock hazards if some of the internal circuitry terminals are touched. (See Sec. 3-3.)

In the demagnetizing process the demagnetizer is kept away from the VCR during the time it is plugged into the wall socket and the button depressed to turn it on. The unit is then brought near the heads.

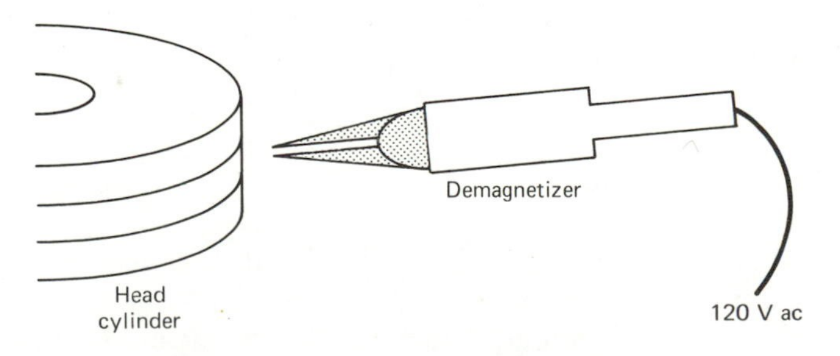

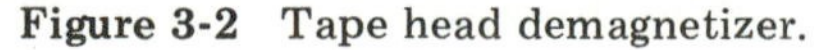
Figure 3-2 Tape head demagnetizer.

In doing this, take precautions not to touch the head with the end terminals of the demagnetizer. This minimizes the danger of scratching the surface of the head. The demagnetizing process consists of bringing the demagnetizer as close to the head as possible and moving it back and forth near the head. Since there is some magnetic pull between the head and the demagnetizer, make sure that a firm grip is maintained to prevent actual contact. With the demagnetizer still turned on, repeat the process for the other heads. When the demagnetizing process is completed, *do not release the on button* until the demagnetizer is over 6 ft (approximately 2 m) away from the VCR. If the demagnetizer button is released while the magnetizer prods are near the heads, it will magnetize the heads rather than demagnetize them. This comes about because the shutting off of the magnet creates a sharp collapsing field which will induce magnetism into any nearby iron or steel objects, including the heads.

3-6. THE PROBLEMS WITH DUST

In all routine maintenance procedures one necessary precaution is the exclusion of dust from the inside of the VCR cabinet. Since the VCR has a number of moving parts as well as sensitive heads, an accumulation of dust can cause considerably more problems than would occur in an audio cassette recorder. An accumulation of dust around the heads can act as an abrasive factor when the heads slide over the tape. Similarly, the entrance of dust into the bearings of pulleys and pinch rollers can hinder free movement and contribute to uneven recording or playback.

To minimize the accumulation of dust in the VCR, the latter should not be kept near hot-air ventilators or open windows. Neither should the VCR be placed in a position where it receives direct sunlight for any length of time. The metal cabinet housing can overheat and damage some of the components. It is advisable to use a dust cover on the VCR to minimize the entrance of dust to the air vent openings of the cabinet. During operation, of course, the dust cover should be removed to minimize heat buildup within the cabinet and chassis.

3-7. PARTS LUBRICATION

When the housing is removed from the VCR for head cleaning and demagnetization purposes, a general inspection of all moving parts should be made. In particular, when heads are cleaned and demagnetized, it is advisable to lubricate those sections which are in continual

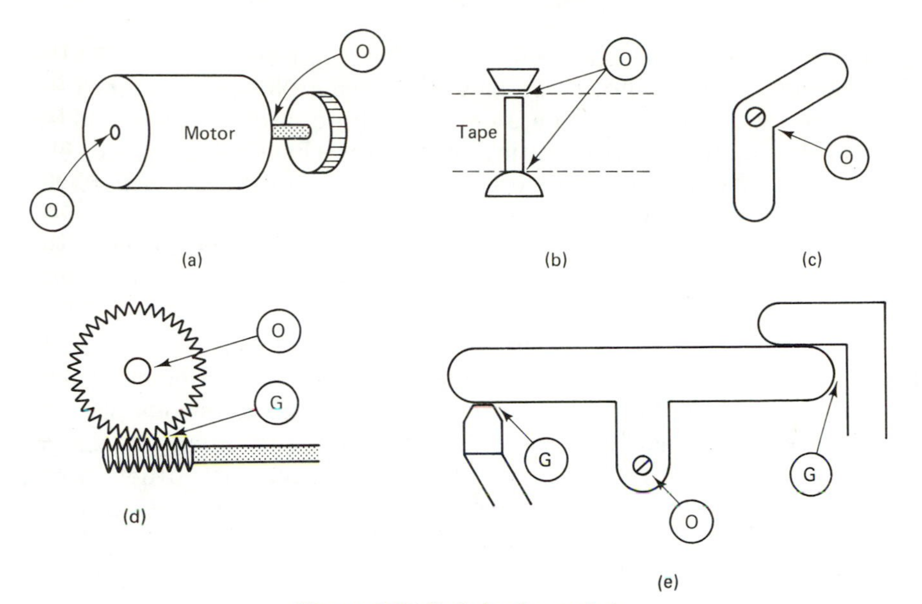

Figure 3-3 Lubrication points.

motion during play or record as opposed to the movement undertaken when the pause control is depressed or when the video cassette is inserted or removed from the unit. Typical of the sections which should be oiled are pulleys, wheels, bearings, and similar sections, as shown in Fig. 3-3.

Motor bearings, as shown in Fig. 3-3(a), should be lubricated using the oil sparingly to avoid oil contamination of the motor coils. Both end bearings should be oiled, though on occasion only one end is exposed. The center moving section of tape guideposts should receive a drop of oil at the top and bottom, as shown in Fig. 3-3(b). Swivel sections, as illustrated in Fig. 3-3(c), should be oiled at the area indicated by the arrow. The letter O in the circle denotes oil, as opposed to the letter G, which indicates that grease is preferred.

As shown in Fig. 3-3(d), ratchet wheels and worm gears are also in need of lubrication, though in this instance grease is preferred as indicated by the G in the circle at the arrow end. Such wheels are used in front-loading tape cassette transport sections and in tape head drum or cylinder drive mechanisms.

A swivel section T bar is illustrated in Fig. 3-3(e). Here the center swivel bearing should be oiled as identified by the arrow and the O inside the circle at the arrow end. For sections that push or slide along such parts, grease is preferred, as identified by the arrows with G inside the circle at the end of the arrows.

Proper oils for the parts shown in Fig. 3-3 are available from electronic jobbers or VCR sales departments. The consistency of the oil should not be so thin that a good film covering of oil is not retained by the bearings, pinch roller, or other similar items. Neither should the oil be too heavy in consistency because it may bind or clog the sections involved. Do not use the penetrating type of oil which has chemicals in it that loosens rusted bolts since this tends to be too thin. It is better to use a pure oil free of any additives.

Use an oil applicator that can be controlled so that an excessive amount of oil is not released. Overoiling can result in the dripping of oil into electronic circuitry or out of the cabinet to the furniture below. After lubrication of wheel bearings, pinch roller bearings, tape guideposts, and so on, it would also be advisable to check for free movement in any push-button controls. A drop or two of oil at various places is advisable even though these parts are not under too much movement. All sections where levers, rods, or extensions touch each other should not be oiled but instead should be greased. Here the composition of the grease should be of medium consistency so that it adheres well and maintains the lubrication for a long period of time.

3-8. BELT TESTING AND REPLACEMENT

In most VCR units some belts are utilized for conveying a rotating motion from one area to another. Although some direct-drive motor units are employed and intermeshed wheels are widely used, some belted portions are still necessary. During routine maintenance procedures it is advisable to check all belts in use. The composition of modern belts is such that they do not tend to become loose when stretched between pulleys. In some instances where a flexible movement is necessary, a slight tension is placed on a particular belt system by having one pulley or wheel in a flexible pivoted mount. A spring attached to the pulley mount then applies enough pull to keep the belt tight on both wheels, as shown in Fig. 3-4. Note where the spring is hooked to the L-shaped tension lever. By moving the spring along the lower arm of the L-shaped section, the tension can be adjusted to correct for a belt that has become loose.

Because of the foregoing factors, it is rare for a belt to become sufficiently slack where it no longer functions properly. Belts in continuous usage, however, will undergo some wear. Hence, when the VCR cabinet has been removed for head cleaning and demagnetization, each belt linkage should be carefully examined and the fingers run along the inside edge to ascertain whether rough spots, cracks, or worn areas are in evidence. If the latter is the case, the belt should be replaced to avoid a possible breakdown in the near future, particularly if an important recording is being made.

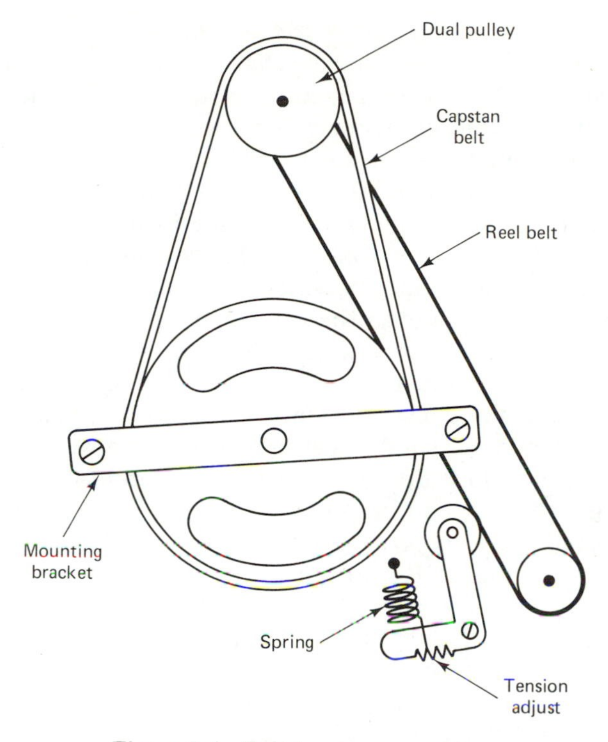

Figure 3-4 Belt tension assembly.

If a belt replacement is necessary, it must be one obtained from the manufacturer of the VCR. Although some small belts are readily available, it is very difficult to get a substitute belt that has the same thickness and width and also the overall length. Usually a substitute belt will not provide optimum performance. In many instances the belt is not as tight as it should be under normal operating conditions, and some slippage may be encountered. Slippage of this sort in some VCR units may upset the servo control sections and cause an automatic shut-off. For the foregoing reasons the best procedure is to obtain the parts number of the belt in question and purchase or order an exact duplicate that will ensure satisfactory operation.

3.9. MISCELLANEOUS MAINTENANCE CHECKS

Once head cleaning, demagnetization, and lubrication have been performed, it is advisable to make several final checks before replacing the cabinet. Initially, the general operation of the entire mechanism should be observed. Insert the ac line into the wall socket and press the on button. Now insert a tape and notice the automatic positioning of the cassette, particularly the front-loading VCR units. Note the manner in which the video head drum assembly rotates and at the same time threads the tape around guideposts and past recording and erase heads and provides the proper slant to the tape as it partially wraps around the recording head assembly. Also observe the movement of the various levers, rods, and metal sections that surround the head cylinder assembly.

Next push the play button and observe the spinning of the cylinder or drum that contains the video recording heads. Note the tape movement in conjunction with the rotating video head drum. Check for any looseness or uneven movement which may indicate some maintenance procedures that must be taken. The tape should move past the heads without squeaking or erratic movement. If these conditions occur, another tape should be tried because a worn tape often exhibits such symptoms. If a new tape does not correct the condition, the heads may not be perfectly clean and should be reinspected and recleaned. Also make sure that the transport spools and rollers move freely in their bearings. Shut off the VCR and remove the tape. Unplug the line cord to minimize shock hazards. Now turn the tape guide rollers and other wheels to make sure all rotate freely without binding. Turn the VCR on again to recheck conditions.

If the tape movement is smooth and all moving parts seem to function properly, the machine can be shut off and the line plug removed before reassembling the unit. When replacing the metal cabinet, make sure that there are no obstructions that hinder replacing the housing in its proper position. Usually any slight problems that arose when removing the cabinet (such as difficulty in disengaging the cabinet from the push button or visual display areas) will occur during cabinet replacement. Care must be taken not to force the cabinet into its proper position to avoid possible breakage or the bending of mounted controls. Check all intercircuit wiring to make sure no extra strands are in the way of cabinet replacement.

Once the cabinet housing has been seated properly, all bolt-mounting holes should coincide with the corresponding openings in the cabinet. When this is the case, all bolts can be put in place. In some instances self-tap-type screws may be utilized in which case the chassis holes that accommodate the screws serve as the holding areas since self-tap screw threads tighten in the holes provided. However, whether screws or bolts are employed, do not insert one and tighten it completely before inserting the others. If one bolt or screw only is tightened to the fullest extent initially, there are occasions when some binding occurs, and it becomes difficult to tighten the remaining bolts or screws to the required extent. It is better to turn the bolts or screws all the way in without applying tightening pressure. Now check the cabinet seating to make sure it is in the proper place and tighten all the bolts or screws necessary.

3-10. ROLLERS AND GUIDEPOSTS

In every VCR there are numerous rollers, tape guideposts, belt pulleys, and similar items that eventually require lubrication or repair. Although the type and placement differ for various models, most have common characteristics that require common oiling, greasing, or replacement techniques. Lack of periodic oiling will eventually impair freedom of movement in the roller or pulley. This could cause binding and excessive tape wear because the tape must rub across the rigid guidepost or roller instead of gliding along by roller post rotations.

Various rollers, pulleys, and guideposts are illustrated in Fig. 3-5. In Fig. 3-5(a) is shown a double-guidepost arrangement with tubular sections that rotate as the tape passes over them. Many such rollers or posts are held in place by an E-shaped spring section resembling a thin open-side washer, as shown in Fig. 3-5(b). Those are pressed into circu-

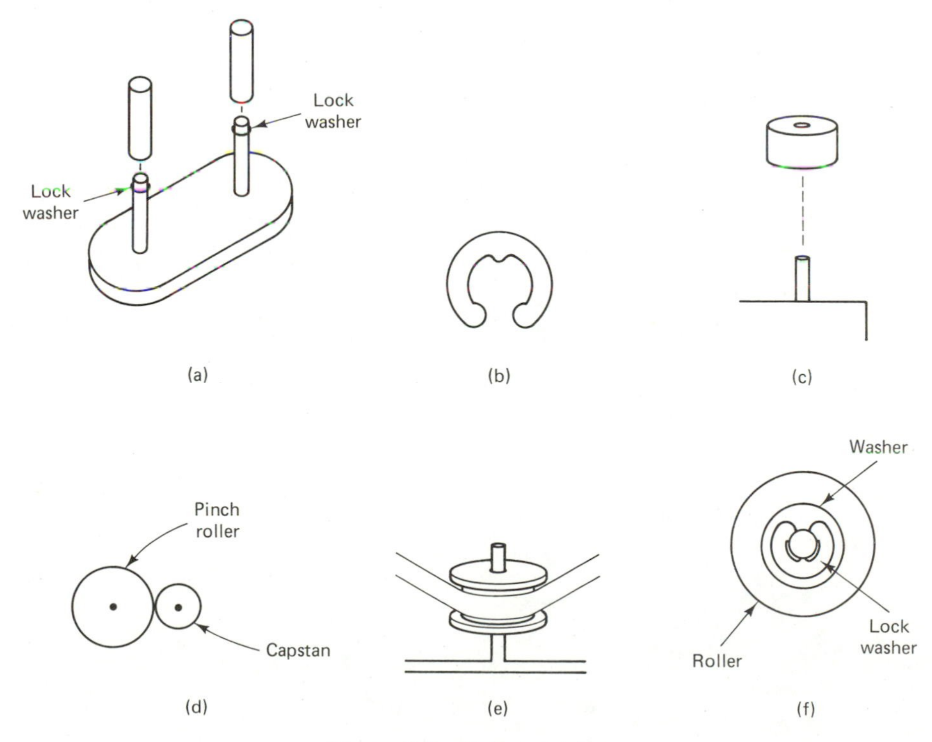

Figure 3-5 Roller details.

lar slits at the top of the mounting post that holds the roller. These E-shaped clamps that lock down the rollers are removed by pulling them out by firmly gripping them with pliers. An awl can also be used to pry the unit from the post. Some tools have been designed for these but are not always available.

Various other rollers and pulleys are also shown in Fig. 3-5. In Fig. 3-5(c) is the rubber pinch roller, with its usage shown in Fig. 3-5(d). The tape winds between the roller and capstan post, and the rotation of the latter moves the tape. In Fig. 3-5(e) is a typical belt pulley, again held down by the E-shaped segment shown in Fig. 3-5(b). A conventional washer is usually present between the roller or pulley and the locking spring washer, as shown in Fig. 3-5(f).

4

Special Symbols

4-1. GENERAL CONSIDERATIONS

For newcomers in servicing or for technicians having a knowledge of basic electronic symbols it is advisable to become acquainted with the digital logic symbols since some of these are usually included in the schematics for VCR units. The logic symbols are somewhat different from the ordinary symbols, such as those shown in Fig. 4-1. Here each symbol represents a specific component, and the physical counterpart is shown so that the component can be recognized. In the logic symbolization each symbol actually designates a complete circuit, and thus the utilization of such logic symbols makes for more compact schematics. The logic symbols also permit immediate recognition of certain circuits used in digital logic. Hence, these symbols more readily identify a circuit and its function than would be the case if the complete circuit were drawn.

The initial illustration of basic symbols will serve as a review for those already familiar with them. For the newcomer in electronics and VCR repairs these symbols represent an important and necessary knowledge that one must have to facilitate any troubleshooting and repair procedures. The symbol in Fig. 4-1(a) represents a resistor, and a representative actual resistor is shown in Fig. 4-1(b). Here the stripes represent color bands, as covered more fully in Appendix E. A resistor can also be shown as a variable one as in Fig. 4-1(c), with the arrow representing the movable arm of the resistor.

The symbol in Fig. 4-1(d) represents a capacitor, and a typical unit

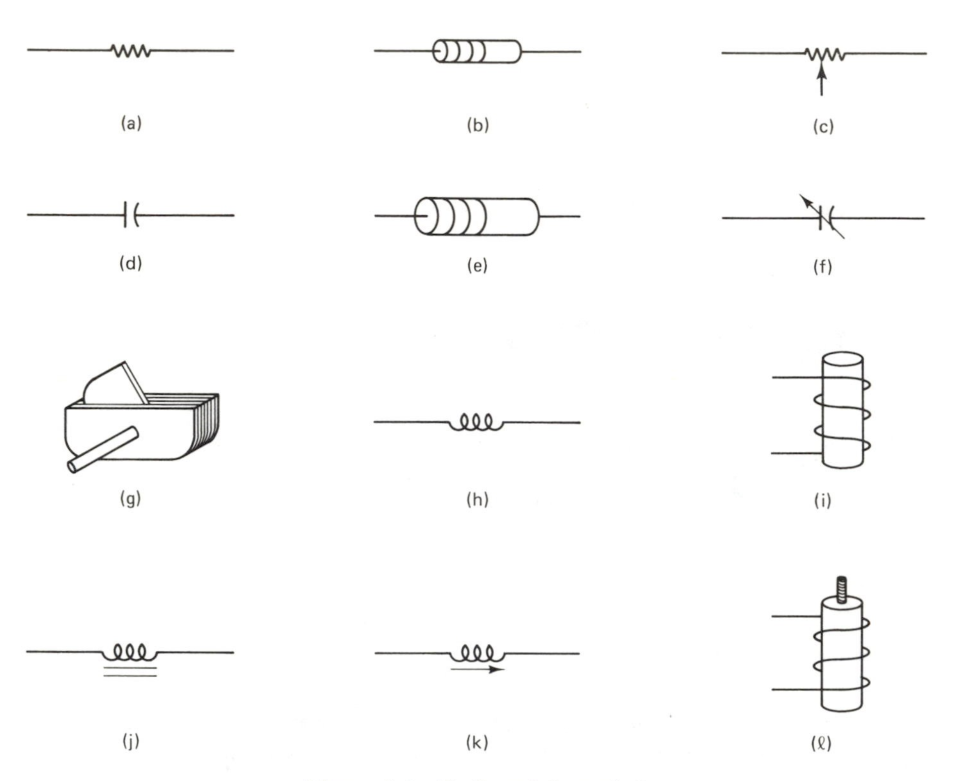

Figure 4-1 R, C, and L symbols.

of this type is illustrated in Fig. 4-1(e). Again, the color bands give ratings, as covered in Appendix E. Reference should also be made to Chapter 7 where the larger-value capacitors are illustrated in conjunction with power supply circuits. Capacitor symbols with arrows through them as shown in Fig. 4-1(f) represent variable capacitors. Such variable types found extensive usage in radios where manual tuning varied the capacitance and selected a new station. Variable capacitors utilize a series of meshing plates, as shown in Fig. 4-1(g). As the tuning knob was turned, the plates either meshed or unmeshed to tune to a lower- or higher-frequency station. In varactor tuning processes such variable capacitors have been replaced by push-button selection (see Fig. 1-11).

A coil (inductor) is shown in Fig. 4-1(h). Such a coil may consist of many windings such as in the primary or the secondary of a transformer, or it may consist of only a few windings for UHF tuning purposes, as shown in Fig. 4-1(i). When a coil has straight lines drawn beside it as in Fig. 4-1(j), it indicates an iron core. If an arrow is shown as in Fig. 4-1(k), it indicates a core that can be moved within the coil and thus alter its inductance for tuning purposes. Such a core may have a slotted area which can be turned to move the core, as shown in Fig. 4-1(l).

Typical diode symbols and representative physical counterparts are shown in Fig. 4-2. In Fig. 4-2(a) is the diode symbol showing the anode and cathode sides. In any diode electron flow is only in one direction, as indicated by the broken-line arrow. In Fig. 4-2(b) is shown a low-power diode with a band imprinted on it to denote the cathode end. A heavy-duty diode (for rectification purposes, as described in Chapter 7) is shown in Fig. 4-2(c). Some diodes may also appear as shown in Fig. 4-2(d) and (e). The varactor diode which has capacitive characteristics is illustrated in Fig. 4-2(f) where either symbol is valid and both have been used extensively in the literature. The zener diode symbols are shown in Fig. 4-2(g), and as with the varactor diode both symbols are found in the literature. Zener diodes are used for voltage stability, as discussed in Chapter 7.

Miscellaneous special diodes are illustrated in Fig. 4-3. In Fig. 4-3(a) is the silicon controlled rectifier, which is a switching diode having an extra lead (gate). A voltage applied to the latter permits conduction through the diode. The letter symbol for such a diode is SCR. A diac is shown in Fig. 4-3(b), and this is also a switching diode, since it does not conduct until a specific breakdown voltage is reached. When a gate element is added to the diac, it becomes a triac, as shown in Fig. 4-3(c). A voltage applied to the gate terminal permits conduction in either direction.

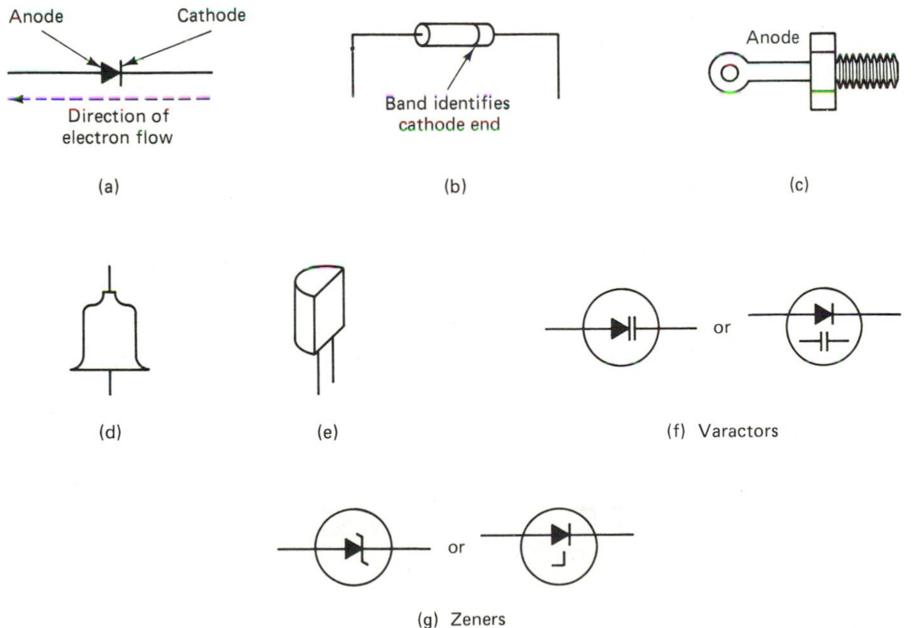

Figure 4-2 Various diode symbols.

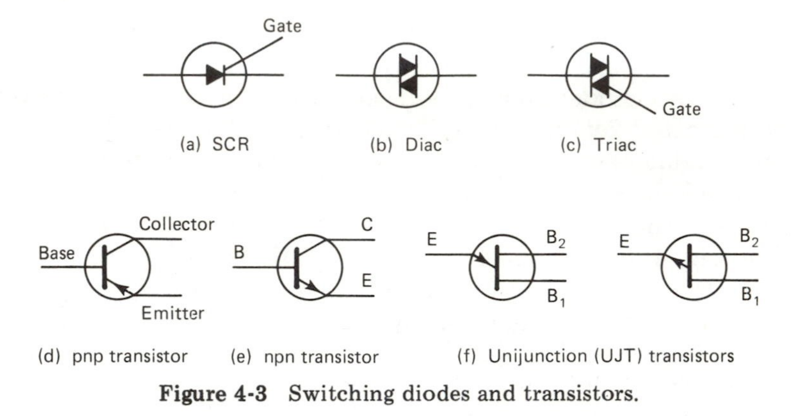

Figure 4-3 Switching diodes and transistors.

Basic transistor types are illustrated in Fig. 4-3(d) and (e). The transistor symbol in Fig. 4-3(d) is known as the *pnp* type since the emitter terminal is normally positive with respect to the base. The type in Fig. 4-3(e) is the *npn* type, and this is distinguished by the arrow pointing out at the emitter. For the *npn* type the circuit polarities are opposite to those of the *pnp* type. In the *npn* type the emitter is negative with respect to the base. Special transistors are shown in Fig. 4-3(f) and (g) and depict the unijunction types. As with basic transistors the two types have opposite voltage polarities in circuits.

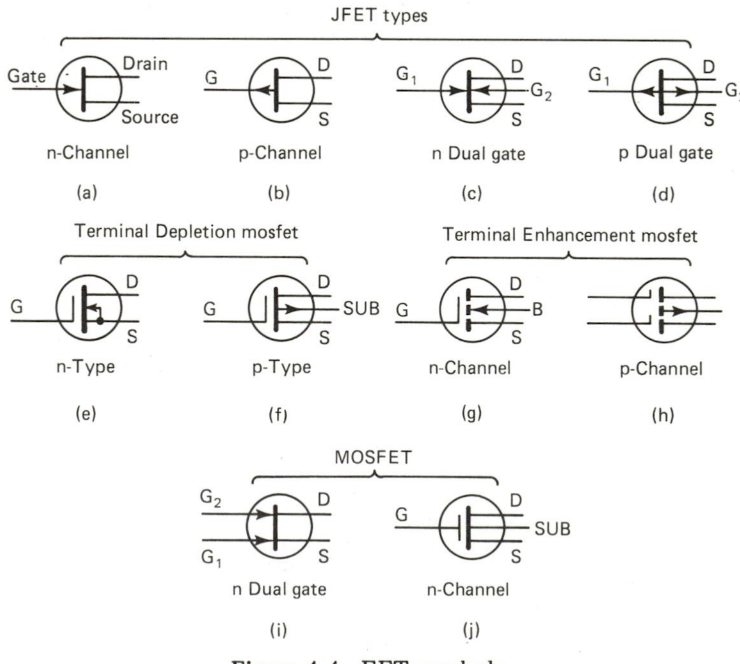

Figure 4-4 FET symbols.

Another type of transistor widely used is that referred to as the *field-effect* transistor (FET). The types illustrated in Fig. 4-4 are representative of the most widely used types. In Fig. 4-4(a) is the *n*-channel FET having gate, drain, and source elements as opposed to the emitter, base, and collector elements used for the basic transistor types. In Fig. 4-4(b) is shown the *p*-channel transistor, which has characteristics similar to that in Fig. 4-4(a) except for the polarity difference in voltage requirements at the elements. Dual-gate types are shown in Fig. 4-4(c) and (d). All these are categorized as *junction field-effect transistors* (JFETs). The other types are known as *metallic-oxide field-effect transistors* (MOSFETs) and are illustrated in Fig. 4-4(e)–(j). The ones in Fig. 4-4(e) and (f) are the *terminal depletion* types of units, while those in Fig. 4-4(g) and (h) are the *terminal enhancement* types. Other representations for the MOSFET are in Fig. 4-4(i) and (j). The dual gate is shown in Fig. 4-4(i), and the one in Fig. 4-4(j) has an element marked SUB that indicates the *substrate* elements (the solid-state foundation slab on which the unit is built).

4-2. LOGIC AMPLIFIER AND OR SYMBOLS

A basic symbol for an amplifier system is a triangle, as shown in Fig. 4-5(a). In this representation it is assumed that there is no phase inversion of the signal from input to output. Thus, if a positive pulse were applied to the input, a positive pulse would also appear at the output except that it would be an amplified version of the input. If the triangle representation has a small circle at its output, as shown in Fig. 4-5(b), it represents an amplifier with a phase inversion capability. Hence, a positive pulse applied to the input would result in the formation of an amplified replica of the input pulse except it would be negative as shown. Similarly, if the input pulse were negative in polarity, the output amplified version would be positive.

A common symbol of the logic circuit family is that referred to as an OR circuit. The designation indicates its function since an OR circuit will combine two *or* more inputs to form a single output, as shown in Fig. 4-5(c). Here, if two pulses were entered at the upper input line and one at the lower, the output would consist of a combination of these as shown. If two inputs are designated as A and B, the OR circuit will have an output for A *or* B *or* both. The symbol in Fig. 4-5(c) may represent a number of actual components. This is illustrated in Fig. 4-5(d) where a group of diodes and a resistor form a three-input OR circuit that develops a common output across the load resistor R_L. For this circuit there is no amplification, and three simultaneous inputs (such as three pulses) would result in only one pulse output. When tim-

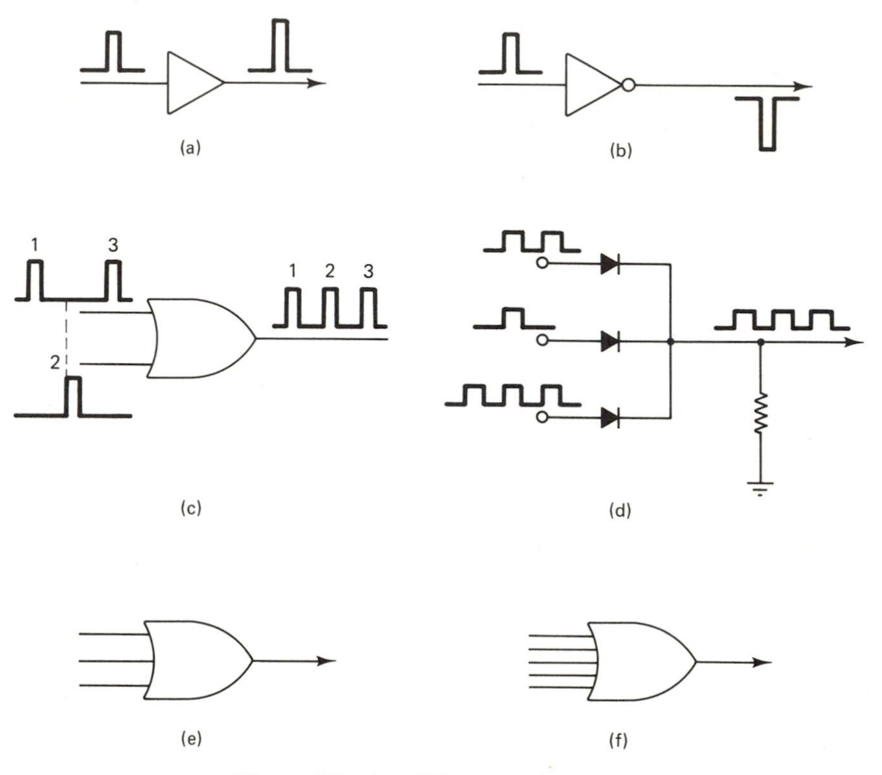

Figure 4-5 Amplifier and OR sections.

ing differences to the input signals prevail, such as shown in Fig. 4-5(d), the output waveform represents the aggregate of the input signals. The symbol for such a three-input OR circuit is as shown in Fig. 4-5(e). An OR circuit can have a number of inputs, as shown in Fig. 4-5(f), where five inputs are present.

4-3. THE NOR CIRCUIT

As with the amplifier shown in Fig. 4-5(b), the inversion symbol can also be used for the OR circuit to convert any output to a phase-inverted representation of the input. Thus, for an OR circuit such as was shown in Fig. 4-5, the addition of a small circle at the output point forms a NOR circuit, as shown in Fig. 4-6(a). This term originates from the concept that a phase inversion means that the output phase is *not* the same as the input. Hence, the full term is NOT-OR, shortened to NOR. Here, if positive pulses were applied to both inputs, a negative pulse output would be obtained. Sometimes letter symbols are used as representative

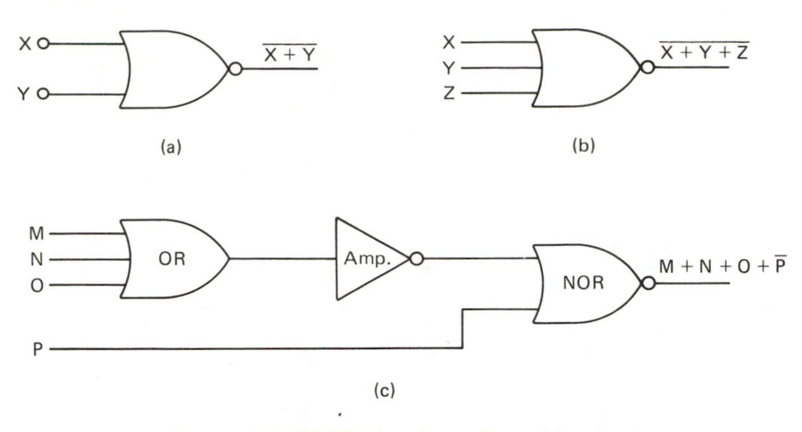

Figure 4-6 NOR circuits and combinations.

input and output sections. Thus, if one input were designated as X and the other as Y, the output from an OR circuit would be $X + Y$. For the NOR circuit, however, the negated $X + Y$ is shown with an overbar to indicate the phase inversion process. As with the OR circuit, the NOR circuit can have several inputs, as shown in Fig. 4-6(b). Here the input consists of X, Y, and Z, and the output is $\overline{X + Y + Z}$. In logic terminology the + sign used between symbolic letters is termed a *logical connective* and does not necessarily indicate the additive process.

All the foregoing logic symbols can be combined in sequence when necessary, as illustrated in Fig 4-6(c). Here a three-input OR circuit feeds an amplifier that inverts the phase of the signals received and in turn feeds one input of a NOR circuit. Thus, if we represent the input as $M + N + O$ for the first OR circuit, this group appears as $\overline{M + N + O}$ in inverted form. This is applied to the output NOR circuit, which also receives the input P. Thus, the output would be $M + N + O + \overline{P}$ because the NOR circuit inverted the signals received from the amplifier inverter and thus restored them to their original form. The P input for the NOR circuit, however, was inverted and appeared as a negative quantity at the output.

4-4. LOGIC AND SYMBOLS

Another logic circuit widely used in electronics and particularly in digital systems is the AND circuit. A transistorized version is shown in Fig. 4-7. In this circuit no output is obtained unless a signal is applied simultaneously to each input. This is termed an AND circuit because an output is obtained only if a signal appears at both the upper and lower inputs. If we designate the inputs as A and B, the logic expression

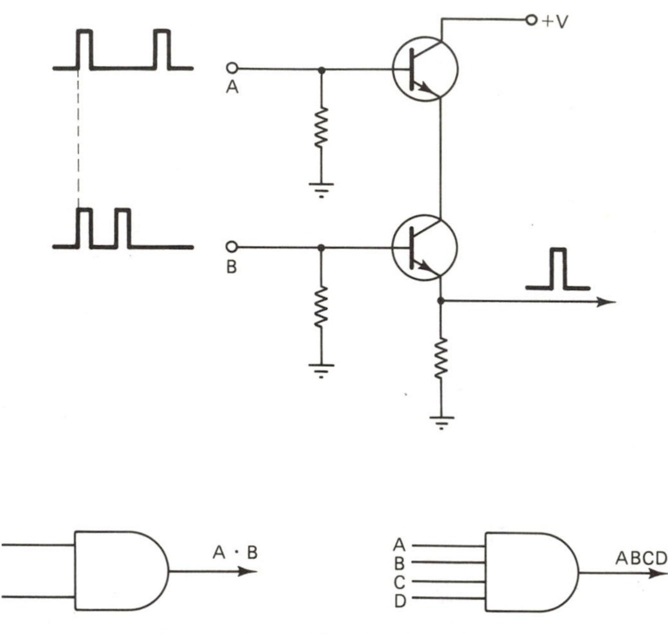

Figure 4-7 AND circuit logic.

would be $A \cdot B$. Note that the AND circuit does not use a + sign as with the OR circuit. Instead, the conventional mathematical multiplication sign (•) is used. Similarly, the multiplication is also indicated when the two letters are written next to each other as AB. Although both these designations imply multiplication, in logic systems the multiplication sign is considered to be a logical connective representative of an AND circuit.

Note the pulse input representation shown in Fig. 4-7. Assuming the rightmost pulses are applied to the inputs of the transistor base element circuits, the first one to be applied would be the upper right pulse to input A. However, at this time there is no coinciding pulse applied to input B, and hence no output is produced. Similarly, the next pulse that enters would be the rightmost pulse for input B. Again, there is no signal input available at this time for the upper transistor; since we have no coincidence, there is no output. When, however, both leftmost pulses are applied simultaneously to the two inputs, an output is obtained from the AND circuit.

Note that the signal output is obtained from the emitter circuit of the lower transistor. Thus, the phase of the output signal is the same as that of the input signals. The symbols for the AND circuit are shown below the schematic in Fig. 4-7. As with the OR circuit, several inputs could be used. A four-input type is shown at the right and the logic output is $ABCD$. The AND circuit is sometimes referred to as a *coincidence circuit* because coinciding signals at the input produce an output. The AND circuit also has switching and gating characteristics that are very useful in VCR circuitry. Several modes may have to be switched such as fast forward, stop, pause, and so on. Aspects of gating and switching are covered more fully in Sec. 4-7.

4-5. LOGIC NAND SYMBOLS

When the output of the two-transistor circuit shown in Fig. 4-7 is obtained from the collector of the upper transistor, a phase-inverted output signal is obtained, as shown in Fig. 4-8. Thus, if coinciding negative pulses are applied to the two inputs, a positive pulse is obtained at the output as shown. Thus, this forms a NAND circuit because the output phase is NOT the same as the input. Consequently, this is a NOT-AND circuit, commonly designated as NAND. Typical symbols for the NAND circuit are shown in the lower part of Fig. 4-8. At the left we have an AB input, which gives us a negated $\overline{AB}$ expressed with the overbar. Similarly, for the three-input NAND circuit shown at the right the output expression is $\overline{ABC}$.

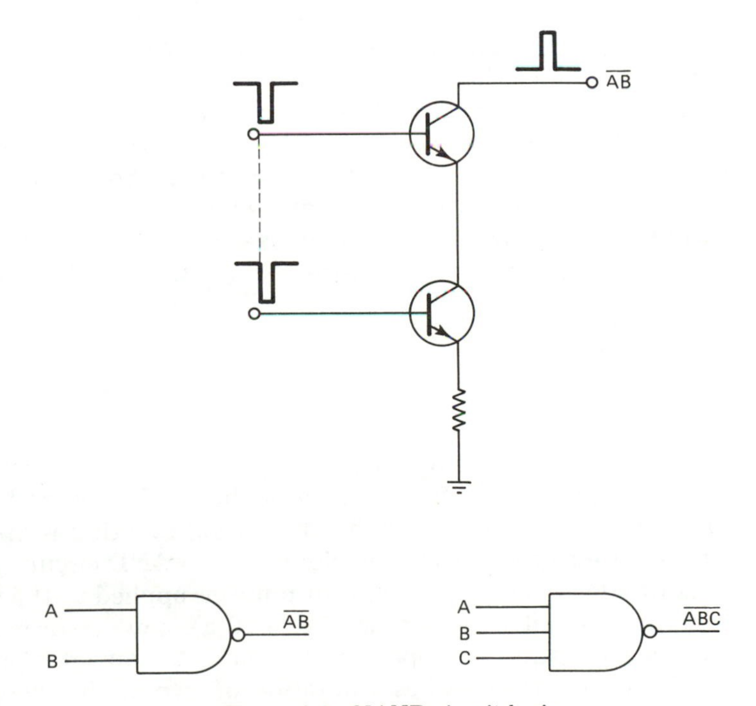

Figure 4-8 NAND circuit logic.

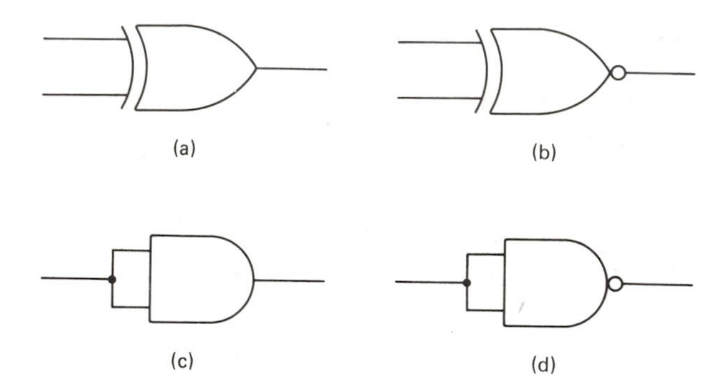

Figure 4-9 Exclusive OR and buffers.

4-6. EXCLUSIVE OR AND BUFFERS

Another symbol of the logic circuit family is that shown in Fig. 4-9(a). This represents the function designated as *exclusive OR*. The latter is an OR circuit utilized in some logic systems and counters. For the symbol shown in Fig. 4-9(b) a negation is present, making it an exclusive NOR circuit. In counting functions it has additive characteristics. The logic expression is $A + B = 0$. In counting circuit structures, the exclusive OR will add without carry, and an additional output line supplies a carry-pulse signal. Although the equivalent circuit function is used on occasion in VCR circuits, the symbol itself is not seen too frequently.

On occasion a circuit is utilized to act as a buffer stage for isolating input and output systems electrically. The symbols representing such buffers are shown in Fig. 4-9(c) and (d). For the symbol in Fig. 4-9(c) the signal undergoes no amplification or phase inversion. For the circuit shown in Fig. 4-9(d), however, a phase-inverted output is obtained. Again this symbol represents a buffer stage that contributes no amplification.

4-7. SWITCHING FUNCTIONS

The logic AND and NAND circuits can be used as electronic switches and thus gate in or gate out signals without the mechanical functions associated with a conventional switch. This gating function is possible because of the necessity for having coincidence in an AND circuit before an output is obtained. Thus, if a series of pulses is applied to the upper input of an AND circuit, as shown in Fig. 4-10(a), any one or more of these pulses can be gated to appear at the output. Thus, if a pulse is entered in the lower line that has a duration of two of the pulses appearing in the upper input, coincidence would be obtained only for the

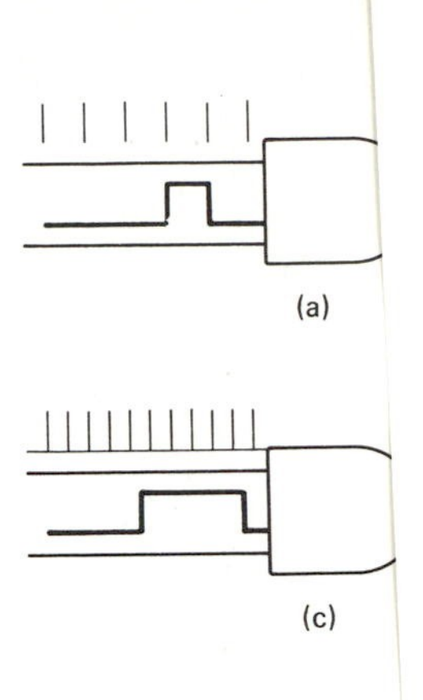

fourth and fifth pu
pulses are switched
is again applied to
pulse train shown
obtained for the m
would be two pulses as shown.

The NAND circuit function can be used similarly as a switch, as shown in Fig. 4-10(c). Here a pulse train is again entered in the upper terminal, and a wide pulse is entered at the lower terminal. Since the wide pulse spans five of the short-duration pulses, the output would consist of five pulses only as shown. Similarly, for the AND circuit shown in Fig. 4-10(d) a video signal is shown entering the upper terminal. Note that the horizontal blanking pulse contains the 3.58-MHz burst signal. A pulse is applied to the lower terminal, and this pulse coincides with the appearance of the burst signal. Hence, the latter only appears at the output. Thus, the AND circuit serves as a periodic switch for gating in the timing signal as required.

4-8. STOP FUNCTION SYSTEM

A typical application of the OR circuit in a VCR system is illustrated in Fig. 4-11. Here, tape movement is sensed to detect a sudden stop such as occurs at the end of the tape during play, record, rewind, and so on. A stop must also be detected in case of tape jamming or for other abnormal reasons. The detection of any stop permits the design of protective measures to protect the motor system and to shut the VCR off immediately. The OR circuit is useful in this application as shown in Fig. 4-11. Here the output from the OR system trips a solenoid and immediately shuts off the power when the tape end is sensed or if the

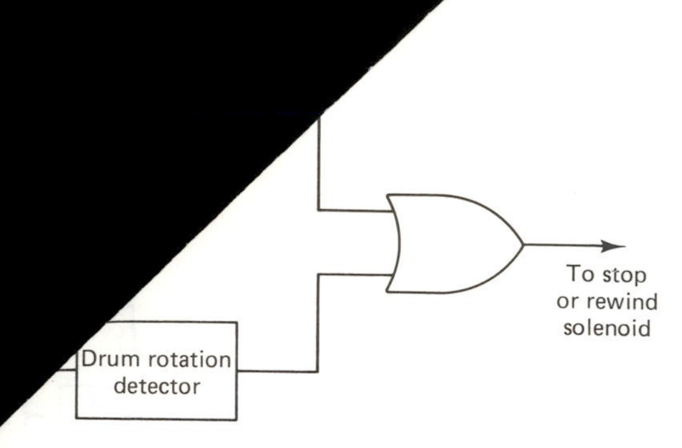

Figure 4-11 Tape end sensing.

ation stops. Tape end detection can be accomplished in sev-
ays. Mechanical sensing can consist of a sensing lever across
h the tape rides, as shown in Fig. 4-12(a). When the tape end occurs,
e tape becomes taut and moves the sensing lever slightly to initiate the stop command. Another method, electronic sensing, is that wherein the end of the tape contains a section consisting of conducting foil. As the tape slides across the tape end sensor, it signifies the end of the tape when the metallic foil shorts the two sensing electrodes and closes a circuit to initiate a solenoid. Instead of stopping the machine at the end of the tape, the detector section can be designed to rewind the spool automatically. Thus, the tape is rewound for playback since in a VCR system the tape is only recorded and played back in one direction as opposed to the two directions utilized with audio cassette recorders.

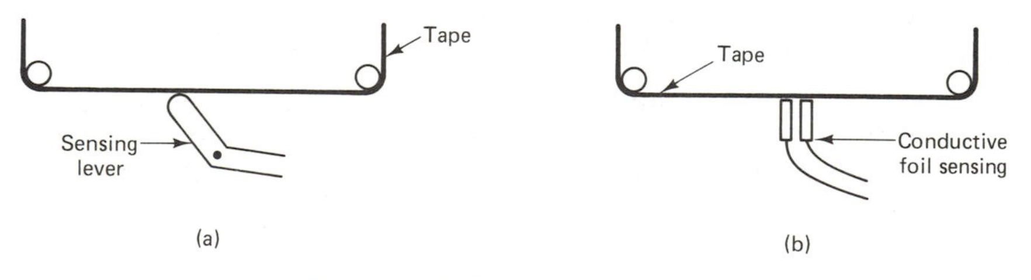

Figure 4-12 End of tape sensing.

4-9. KEYED AGC LOGIC

Utilization of an AND circuit in conjunction with an amplifier circuit is shown in Fig. 4-13. This is the basic circuit of a keyed automatic gain control system. The output voltage developed by this circuit has an amplitude related to that of the incoming video signals. This output voltage is the AGC signal applied to the gain control sections of the tuner and IF amplifiers. By changing the transistor bias, the amplifier gain can be increased or decreased. Consequently, this system maintains

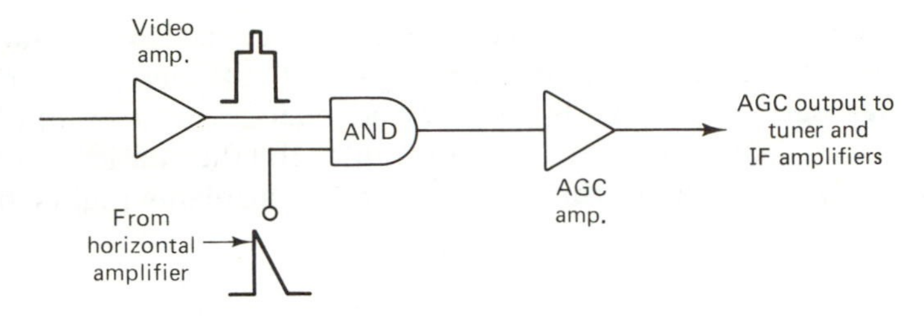

Figure 4-13 System outline.

a constant video signal level as set by the manual contrast control in a VCR monitor or TV receiver. If the incoming signal amplitude increases and thus would have a tendency to overload the monitors circuits, a compensating AGC output lowers the gain of the tuner and IF stages. Thus, the AGC maintains an automatic control of video signal gain.

As shown in Fig. 4-13, the video signal is obtained from a video amplifier and applied to one input of an AND circuit. Since this is a coincidence circuit, the video signal information will not be transferred to the AGC amplifier unless a signal is also applied to the lower terminal of the AND circuit. A pulse is obtained from the horizontal amplifier and applied to the lower terminal as shown. The pulse has a repetition rate of 15,734 pps (the horizontal rate for color receivers). Thus, the coincidence of the horizontal pulse at a time when the video signal contains the horizontal blanking pulse provides the necessary switching, and the AND circuit therefore produces an output that is applied to the AGC amplifier.

When the AND circuit produces the necessary output, it initiates the automatic gain control potential that is applied to the tuner and IF amplifier stages. If the incoming signal declines, the blanking pulse amplitude and sync will be lower, and hence the output from the AND circuit is reduced accordingly. The correction potential will then be such that the gain of the tuner and IF amplifiers is increased to compensate for the decline in the television signal amplitude. Similarly, if the incoming horizontal blanking amplitude rises as the incoming video signal increases, the resultant AGC potential would cause the RF and IF stages to have less gain, to compensate for the video signal rise. Since the AGC signal is, in essence, keyed in only during horizontal blanking, it is less susceptible to noise pulses which might interfere with reception.

When noise pulses are picked up and arrive in combination with the video signals, the noise signals will have no influence on the AGC system. This occurs because the AND circuit only produces an output during the horizontal blanking portion of the complete video signal. Since the AND gate is inoperative during the video signal sections (that contain the noise signals), the noise signals have no influence on the

AGC system. Also, since the AGC signal is related to the horizontal sweep frequency, the circuit time constant is sufficiently short so the system has decreased sensitivity to rapid changes of the input signal. Thus, such an AGC system tends to minimize the fluctuations in contrast on the television screen that are caused by airplane flights over the reception area.

4-10. FLIP-FLOP LOGIC

A flip-flop circuit is widely used in all branches of digital circuitry, including the speed control systems of video cassette recorders. The flip-flop is a switchable dual-state circuit that produces an output only when an input signal is applied. Hence, the flip-flop circuit is not a free-running signal generator or oscillator that produces a continuous signal. In the basic circuit either a pair of NOR gates can be used, as shown in Fig. 4-14(a), or a pair of NAND gates, as shown in Fig. 4-14(b). The input lines are marked R for *reset* and S for *set*, while the output lines are designated as Q for a logic 1 output and $\overline{Q}$ for a logic 0 output.

If the flip-flop is in the state where a logic 1 appears at the Q output, a logic 0 appears at the $\overline{Q}$ output. If a logic 1 input signal is now applied to the set (S) line, there will be no change of state since the flip-flop is already in the set position. When, however, a logic 1 signal is applied to the R input, a negative output appears at the Q line. The signal at the Q line is fed into the S input, and the logic inverts and appears as a logic 1 at the $\overline{Q}$ output. This is the second state of the flip-flop. Now an S input signal produces an inverted output at $\overline{Q}$, and the output of this signal is applied to the upper NOR circuit and inverted at the output to produce a logic 1 at the Q line.

Thus, as with all flip-flop circuits, a set input triggers the flip-flop to produce a logic 1 from the Q line. When a reset signal is applied, it causes the flip-flop to revert back to its original state at which time a logic 0 appears at the Q line. Such a circuit is therefore useful for

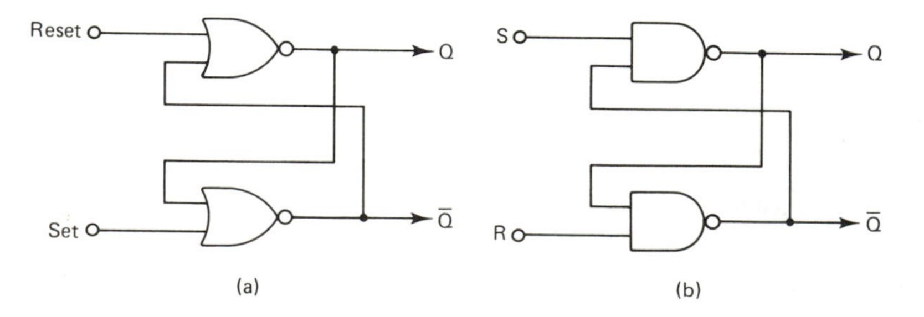

Figure 4-14 NOR and NAND flip-flop circuits.

pulse repetition division, since one output is obtained for every two input pulses. Similarly, it produces a trigger pulse or switching pulse for other gating needs as required.

A similar flip-flop can be formed with NAND circuits as shown in Fig. 4-14(b). The basic function is similar to that described for the OR circuit flip-flop except that to obtain an output signal coincidence must prevail at the inputs of either NAND circuit. The logic for the circuit in Fig. 4-14(b) is the same as for that in Fig. 4-14(a). A set-pulse input produces a logic 1 at Q and a logic 0 at $\overline{Q}$. A signal applied to the reset line produces an inverted signal (logic 0) at the Q line and a logic 1 at $\overline{Q}$. Thus, when a set pulse is applied, followed by a reset, the output from the Q line is a single logic 1 pulse.

The basic flip-flop circuits shown in Fig. 4-14 are usually combined with other logic circuits for obtaining optimum results. A typical representation is as shown in Fig. 4-15(a) where two additional NAND circuits are used to feed the set and reset lines of the original flip-flop. For convenience in reference, each NAND circuit is numbered in the symbol for every NAND circuit, as shown in Fig. 4-15. In this system two inputs of NAND gates 1 and 2 are linked together to form a common timing-pulse input. If the timing pulses are at a fixed repetition rate, they will appear in inputs of the first NAND circuit as well as the

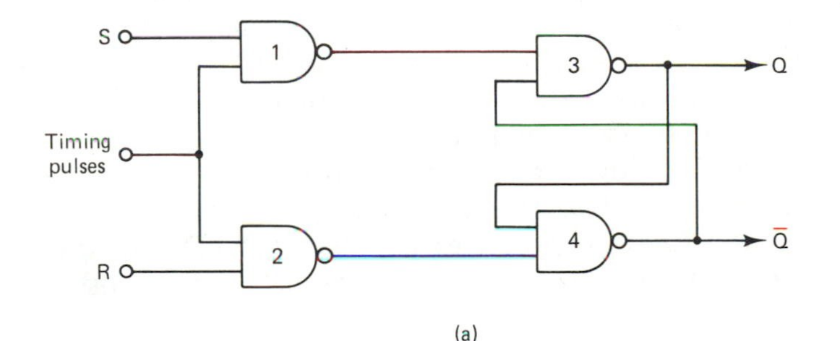

(a)

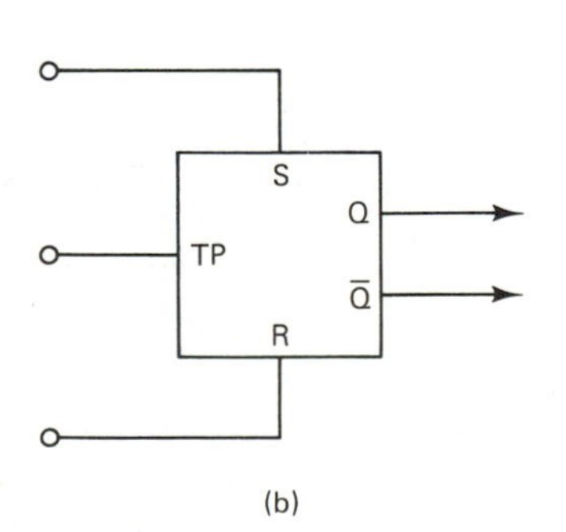

(b)

Figure 4-15 Timed (clocked) flip-flop.

second as shown. Thus, the application of the timing pulses places both these NAND circuits (1 and 2) in readiness for gating in either the S or R signal as needed. If a set function is required, a logic 1 pulse is applied to the S input of NAND gate 1, and coincidence then prevails in conjunction with the timing pulse. Hence, a negated output is produced and applied to NAND gate 3 of the flip-flop circuit. When a reset function is needed, a logic 1 signal input to NAND gate 2 provides coincidence with the timing-pulse input, and the NAND gate becomes an open gate for passage of logic 1. The output from both NAND gates 1 and 2 is negated. Hence, for NAND gate 3 to become operative, a coinciding negative signal must appear at the other input. Consequently, the output from NAND gate 3 again is inverted and represents a logic 1 output on the Q line. This satisfies the set function for a pulse input at the S line. Sometimes such a flip-flop is symbolized as shown in Fig. 4-15(b). The timing pulses are entered in the line shown. The S and R lines are indicated at the top and bottom, with a Q and $\overline{Q}$ appearing at the right.

4-11. PAUSE CONTROL

A typical application of the flip-flop circuit discussed in Sec. 4-10 is the electronic switching in a VCR pause control system. With the pause control function, a push of the pause button should initiate a recording or playback phase freeze for a selected time period. When the pause button is again depressed, the freeze is lifted, and recording or playback continues. The flip-flop circuit is useful because it executes a different function for each of two pushes of the pause button. Thus, for the initial push the flip-flop produces a pulse that initiates the freeze of recording or playback. On the second pause button push, a reverse-

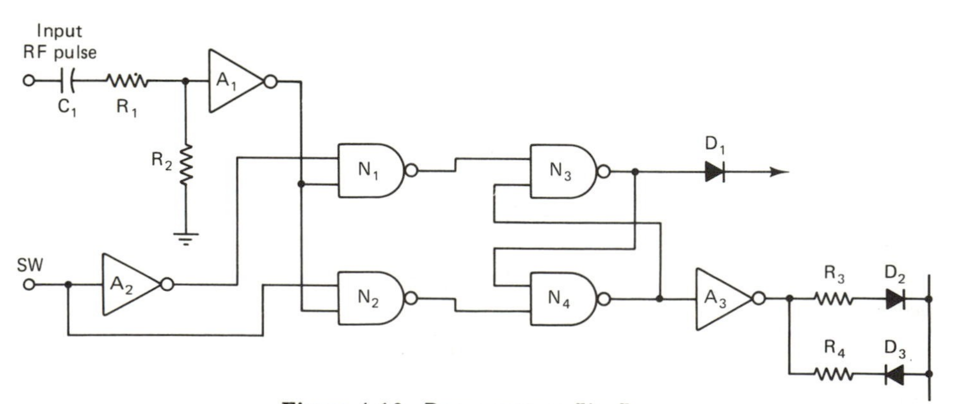

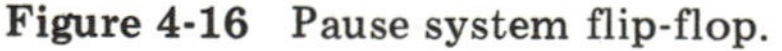

Figure 4-16 Pause system flip-flop.

polarity pulse is produced that is directed to the circuits that cause a reversal of the pause mode.

The basic units of the switch system are shown in Fig. 4-16. A short time-constant differentiating circuit is formed by C_1, R_1, and R_2. This circuit produces a sharp spike-type waveform and forms the equivalent of the timing-pulse input discussed in Sec. 4-10. The pulse is amplified by A_1, the amplifier-phase inverter. The *SW* input terminal is the signal input initiated by pressing the pause control button. In conjunction with the RF spike signal, this signal provides the necessary coincidence in the NAND circuits (N_1 and N_2). The latter feed the flip-flop NAND circuits (N_2 and N_3). Diode D_1 applies a switch signal to the reel motor control circuits, and diode D_3 regulates the tape stop function. When the pause button is again depressed, diode D_2 regulates the tape run function.

5

The VHS Recording System

5-1. TAPE RECORDING BASICS

For an understanding of the functional aspects and recording methods utilized in the VHS video cassette recording system (or the Beta in Chapter 6), a basic knowledge of magnetic recording principles is a necessary foundation for comprehending the more involved aspects of video recording. Essentially the recording of audio on a cassette tape is a simple procedure compared to the recording of video because of the much wider frequency span of video signals and the necessity for recording the accompanying sound. With an audio cassette the standard speed is 4.5 cm/sec, and high-fidelity sound can be recorded with a frequency span of approximately 20 to 20,000 Hz. The earlier magnetic oxide tape coatings were incapable of good frequency response at the high end of the audio spectrum at the standard tape speed, but as tape quality improved, particularly with the advent of chromium-dioxide tapes, the frequency response for audio became fairly uniform with a minimum of attenuation at the high end. For longer play, a thinner tape material is used, and audio cassettes capable of playing 60 min/side are available, though the standard play is 30 min/side.

In all tape recordings the magnetism principle is used. The units employed for recording and playing back the audio material are called *heads*. The tape, with the magnetic materials embedded on one side, passes over two heads, as shown in Fig. 5-1(a). The erase head shown at the left uses a high-frequency ac signal termed *bias* to eliminate any previous recording that may be on the tape. This minimizes interference

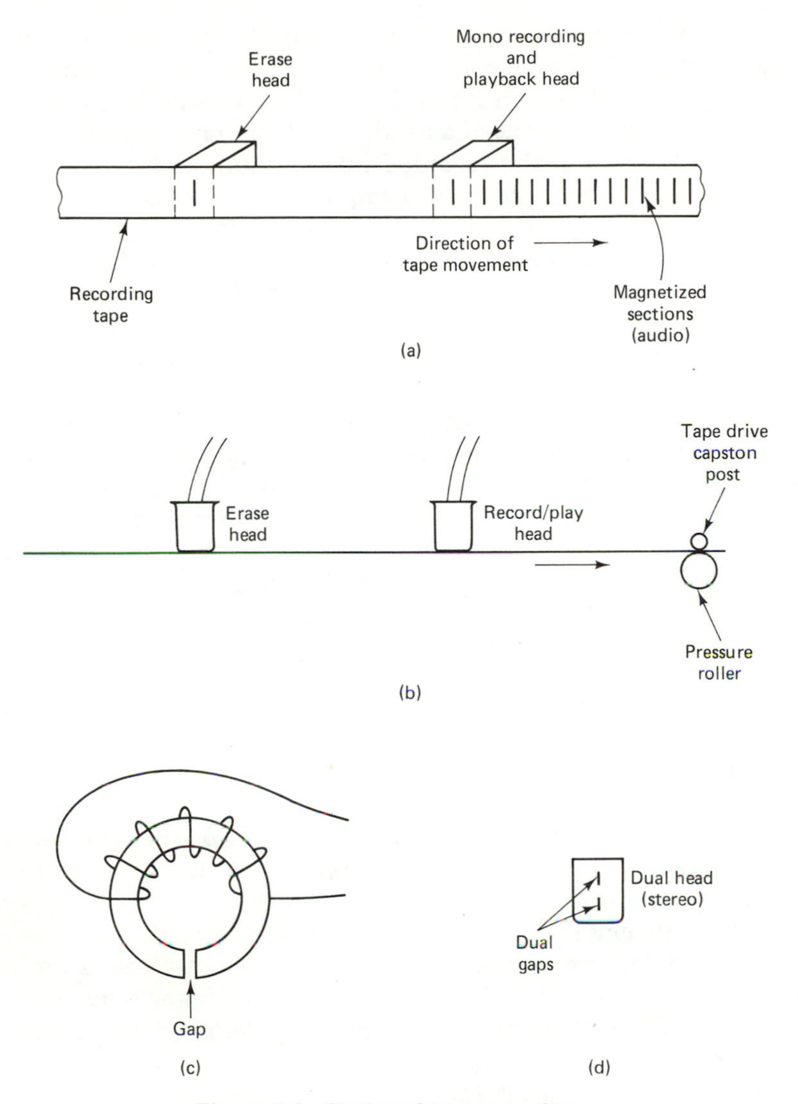

Figure 5-1 Basics of tape recording.

that might occur if prerecorded tapes are used for rerecording purposes. During playback the erase head is switched off automatically. The view in Fig. 5-1(a) is a front view of the tape and heads, with the heads in back of the tape making contact with the magnetic side. The top view is shown in Fig. 5-1(b). Appropriate guides and pinch rollers are used to transport the tape. As shown, the recording head is also used for playback by appropriate circuit switching.

Essentially, a head consists of a coil wrapped around a metallic core that has an extremely small space (gap) in the front as shown. For stereo, dual heads are encased in a single housing (stacked) with the essential gaps as shown in Fig. 5-1(d). When a magnetized segment of tape passes along the front of a head, it induces impulses in the coil and core corresponding to the recorded information. Similarly, if information is applied to the head, it will implant magnetic segments onto the tape that also correspond to the audio information being recorded.

The same basic magnetic principle is utilized in the recording of video information, but a much more elaborate system is necessary because of the wide span of the video signal. (See Fig. 1-9.) Since much higher frequencies are involved than found in the audio spectrum, special measures must be taken to imprint such a wide bandpass magnetically on tape. Consequently, not only is a wider tape utilized, but a recording head movement must also be initiated which in combination with the tape movement permits the recording of the comparatively enormous span of signal frequencies involved in video recording procedures.

To initiate the recording or playback processes, many mechanical devices are involved to move the tape forward, to rewind as necessary, and to permit a fast-forward operational mode. In addition, motors, belts, ratchets, and other necessities are present, as detailed in subsequent sections.

5-2. VHS TAPE SCAN SYSTEM

In the audio cassette systems discussed in Sec. 5-1, the tape heads have a narrow vertical gap in their center area that faces the tape. The gap is made as narrow as possible since it is the focal point of tape recording. If a high-frequency signal is being played or recorded, a narrow gap is desirable to minimize overlapping of high-frequency signal information. In all tape recording procedures the handling of high-frequency signals without loss presents the most challenging problem in design. In addition to a narrow gap, the tape speed as it slides past the heads is also an important factor. The higher the speed, the more readily do we achieve low-loss high-frequency signal recording and playback. In the early days of tape recording the magnetic material used on the tapes also contributed to losses in high-frequency signal recording. Subsequently, however, superior magnetic materials were utilized that increased the frequency span considerably. Where earlier reel-to-reel tape recorders required a tape speed of 10 cm/sec for good high-fidelity recordings, modern cassette machines produce high-frequency and high-fidelity recordings with a tape speed of only 4.5 cm/sec.

In video recording the frequency range of the signal extends to 4 MHz instead of the usual 20 kHz for audio. Thus, there is a need for a range 200 times that for good audio. If we were to increase tape speed in order to obtain a 4-MHz span, it would require a prohibitive amount of tape to record an hour's program. Similarly, the gap cannot be narrowed sufficiently to achieve the higher-frequency response, though extremely narrow gaps are already in use. The VHS system, for instance, uses a tape head gap of only 0.3 μm. However, to achieve the recording of video information, the method of scan had to be changed radically by initiating a rapid tape head spin in conjunction with the tape movement. The basic principle is shown in Fig. 5-2.

As shown in Fig. 5-2(a), a rotating cylinder is used with a direction of spin as shown. Note that the tape wraps around the cylinder, as more clearly illustrated in Fig. 5-2(b). The cylinder contains two tape heads exactly opposite each other as shown. Note also that the tape is placed across the cylinder, diagonally held in place by two guides as shown in Fig. 5-2(b). The cylinder rotates at 1800 rpm, and the two heads scan

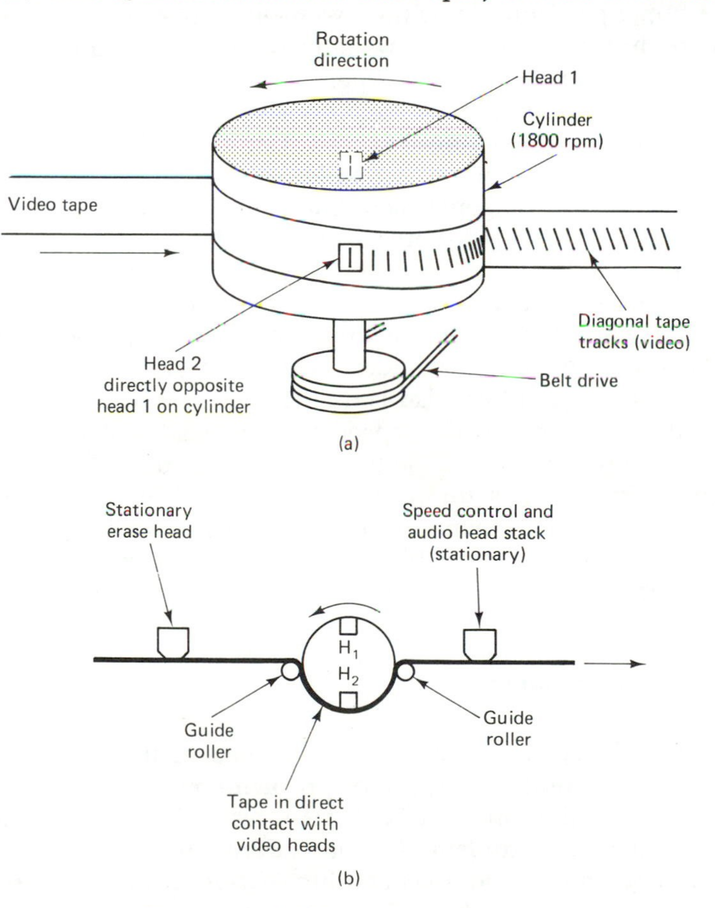

Figure 5-2 VHS helical scan system.

the tape as the tape moves across the cylinder. The result is that though the tape only moves at approximately 2 cm/sec the combination of the rotating cylinder and tape movement gives a *relative speed* of 580 cm/sec (5.8 m/sec). The tape slant across the cylinder plus the increased relative tape speed form diagonal tracks on the tape during recording, and the system accommodates the complete frequency span of the video signal.

Although there are some similarities between the Beta and the VHS systems, they are not compatible because of differences in recording techniques. In the Beta system a drum is used instead of a cylinder, though both have identical speeds of 1800 rpm. For the Beta an 0.6-μm gap is used instead of the 0.3-μm gap for VHS. In both systems the tape speed can be altered for a longer play than standard. With a slower speed there is some degrading of picture and sound quality (see Sec. 3-4).

In Fig. 5-2(a) the lower part of the cylinder assembly is stationary, but the upper part containing the two heads is the rotating section. Several more heads can be used to improve auxiliary functions such as freeze frame or reverse scan. When two heads are used, each head records or plays back one TV field consisting of 262.5 lines. (See Sec. 1-2, Fig. 1-7.) Since each head records one field, the combined recordings make up the 525 lines of a complete video frame.

The slanted or diagonal recording is performed by the helical scan system, and the tape guides must maintain the tape in the proper slant across the cylinder to engage the two heads during the process. As shown in Fig. 5-2(b), the erase head is stationary since its whole function is to make sure that any recording on the tape is erased before the tape reaches the recording cylinder. The erase head circuit is disengaged during playback. Another head is used for audio recording and pickup. This head is combined with a speed control head that records control pulses on the outer track of the tape during recording. During playback the control head senses the prerecorded control pulses and thus regulates the speed in the manner more fully discussed later. Usually the speed control and audio heads are stacked and, as with the erase head, are stationary.

5-3. VHS SOUND FIDELITY

Good high-fidelity sound quality is obtained in the VHF and (Beta) VCR in modern units. Hence, any problems in achieving good sound quality in recording and playback requires a thorough search procedure to localize the problem. The procedure would involve a thorough cleaning of the heads and a general lubrication of moving parts such as pinch rollers, guide wheels, and so on, as discussed in various sections of this text. When, however, some uneven sound quality is present when operating early VCR models, the wow and flutter may well be caused by the inherent poorer sound quality present in these earlier units. Hence, though the usual measures of head cleaning and parts oiling can be undertaken to achieve maximum quality, the end results may still be unsatisfactory.

The problems associated with earlier VCR units involve the handicap imposed by the tape speed and sound track used. As shown in Fig. 5-3(a), the VCR tape is 12.7 mm (½ in.) wide, and the bulk of the space is allotted to the video signals, with only a narrow 1-mm strip along the top used for sound. This track is recorded with a stationary head unlike the video track. Since the tape speed is approximately 2 cm/sec (compared to 4.5 cm/sec for audio cassettes), the sound quality suffers. The slow speed is also conducive to more noticeable volume fluctuations.

With the advent of stereo in the VHS system (1979) the 1-mm sound strip was divided into two 0.35-mm sections, with an 0.3-mm space between them, as shown in Fig. 5-3(b). The right and left stereo channels were carried by the 0.35-mm sections, and the 0.3-mm guardband in the center minimized cross talk between the channels of stereo. The actual sound quality was not improved with this system, but the

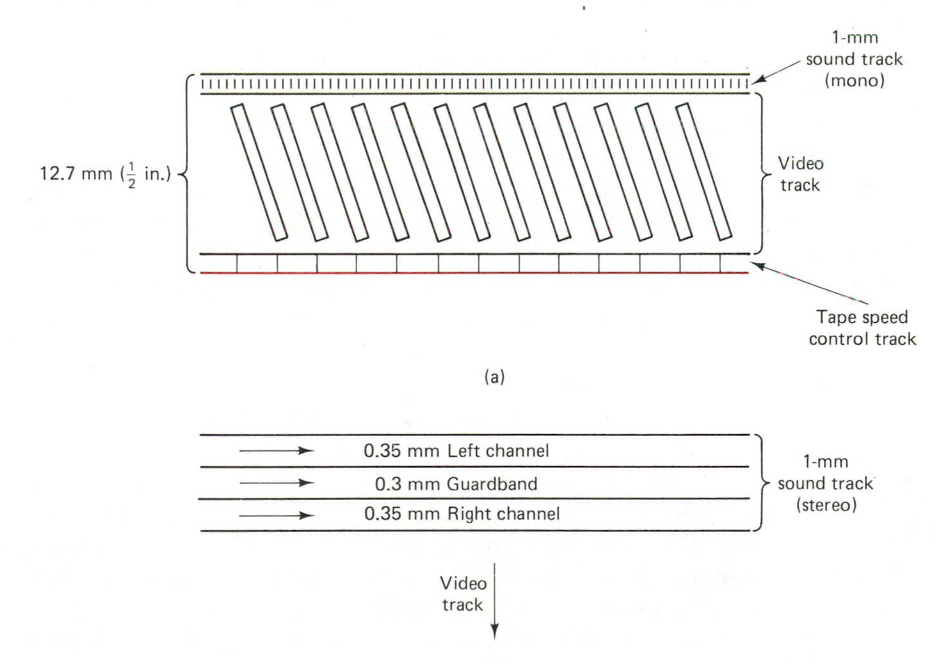

Figure 5-3 Tape tracks for video and audio.

stereo was a good selling point. Best results were obtained by using the highest recording speed rather than the extended play EP or even the LP.

Subsequently (1984), the stereo system was improved considerably by the addition of two separate heads to the cylinder. These additional heads were used to superimpose the audio on the same track as the video, thus taking advantage of the increased relative speed achieved by the combination of the rotating cylinder and the tape movement. Thus, when the recordings are played through a good stereo system, excellent stereo quality is achieved. The original upper audio stereo band is still retained, thus making the system compatible with the older tapes. For these, however, the high-fidelity advantages achieved by the dual mixing of video and sound on the video track are lost. (See Sec. 12-2).

5-4. SERVO CONTROL SYSTEMS

Because the video information has a wide frequency range and the recording process involves a rotating cylinder combined with tape movement, a number of factors are involved that necessitate strict control of the cylinders, the tape speed, and so on. Such control is required to synchronize the heads during playback so that the picture information will be played back with the same fidelity as recorded. Lack of synchronization to even a small degree can result in poor picture quality as well as interfering streaks across the picture.

The method of control is a servo system that senses the rotation of the cylinder and compares it to the pulses recorded on the control track. If no coincidence is achieved, correction is immediately made. Two methods of servo control are generally used in VCRs. One system is the servo control of the rotational cylinder speed. The other is the capstan tape drive servo control. In the cylinder servo control the rotational speed is compared to the control pulse rate and the speed altered if incorrect. For the capstan control, the loss of synchronization is corrected by a speed change of tape, thus altering the number of pulses per second on the control track of the tape.

As mentioned earlier, the rotating heads each scan one field of the television picture frame (1/60 sec). The timing is such that the two fields traced on the screen interlace to form 525 scan lines per frame, or 262.5 lines per field. Since the cylinder rotates at 1800 rpm, the rotation in 1 sec would be 1800/60, which equals 30 Hz. (Here we are using the hertz as reference to *revolutions per second* rather than the standard *cycles per second* denoted by the hertz reference.)

Both the VHS and Beta machines use servo systems for precise control of tape speeds or cylinder rotation. The control track pulse train is formed and precisely timed by the vertical synchronization pulses that accompany the incoming video signals. Essentially these sync pulses have a repetition rate virtually the same as the ac power-line frequency, though a slight difference is present for the vertical scan frequency of color video compared to the black-and-white picture. (As shown in Appendix C, both the horizontal and vertical sync frequencies differ slightly for color and black-and-white pictures.)

The 60-Hz sync pulse train is applied to an electronic circuit that divides the frequency in half and produces a 30-Hz pulse train. The latter is sometimes termed the control (CTL) signal. It is the latter that is recorded on the CTL track illustrated in Fig. 5-3.

The basic principles of a typical servo system are illustrated in Fig. 5-4. During recording the synchronizing pulses are impressed on the control track illustrated earlier. During playback the pulses on the control track are compared to the pulses picked up by the iron-core coil shown in Fig. 5-4. The latter senses the rotational speed of the cylinder because the small magnet embedded in the lower section generates a pulse each time it passes over the coil. Thus, the coil senses the speed of the cylinder and transfers it to a circuit that compares the 30 Hz of the cylinder rotation with the 30-Hz pulses received from the control track. If any variation in cylinder speed is sensed, a correction signal is developed and applied to the speed control section. The latter can then be

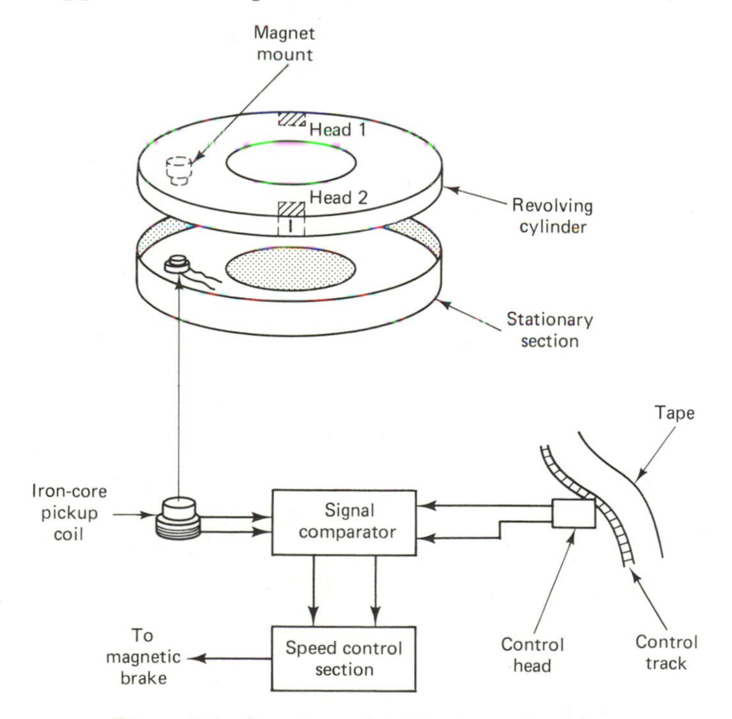

Figure 5-4 Speed synchronization and control.

used to increase or decrease the speed of either the tape or the revolutions of the cylinder to a sufficient degree to achieve the necessary correction. During recording the same sections are utilized in reverse order.

5-5. TRACKING CONTROL

A tracking control is essential in achieving true compatibility with tapes made on other machines. Even among VCR units of the same VHS system, a commercial VHS tape containing a complete motion picture or the tape from a neighbor also using a VHS unit may not give good results because of slight timing differences during recording and playback. To avoid such incompatibility on recorders of the same make or type, a tracking control must be present. With the latter the synchronization pulses in the control track are utilized to make a slight correction so that true control is regained when using a different tape. The tracking control functions only during playback and usually has a standard or central setting marked. The control should be left in this position and changed only when playing a tape recorded by another VCR. After playback the tracking control should be reset to its normal position.

In many VCRs the tracking control consists of a thumb wheel, as shown for the VCR illustrated in Fig. 5-5. In some units the tracking

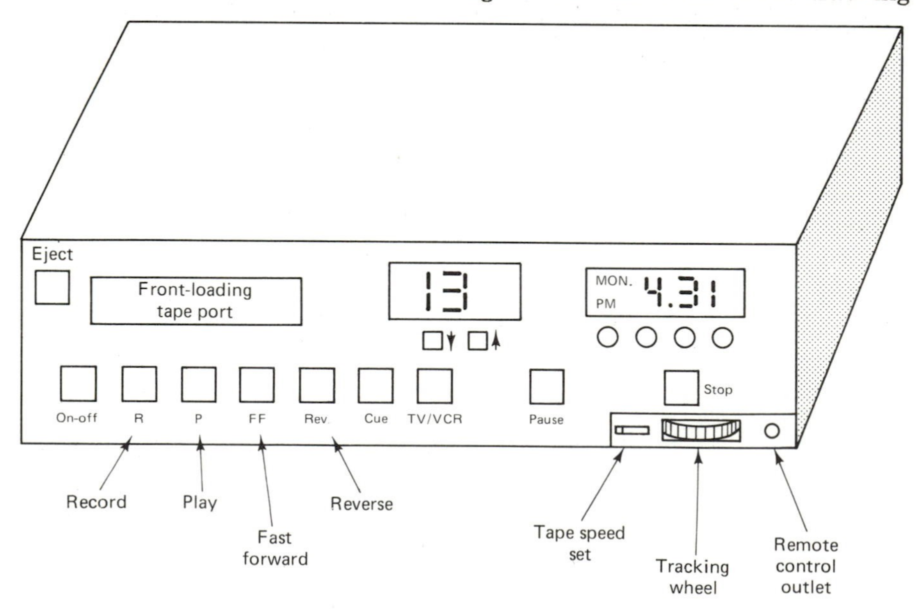

Figure 5-5 Typical control switches.

control may be a conventional knob. For the one illustrated, the tracking wheel is situated at the lower right of the front panel, though other manufacturers may place it in a different position. (See also Fig. 6-5.) The other controls illustrated in Fig. 5-5 are representative, but the exact placement in the various VCR models may vary considerably. Another front-panel configuration is also shown later in Fig. 6-5.

The need for adjustment of the tracking control is indicated when the picture quality is poor and when black streaks appear across the screen during playback. The tracking control is most effective for tape recording at the standard speed (SP). The control is less effective if the recorded material uses a slower speed such as EP. Unless the tape being played is particularly different in characteristics, the signs of mismatched tape speeds can usually be corrected to some degree with the tracking control.

5-6. FREQUENCY-MODULATED VIDEO

Although the rotating heads and the moving tape permit the recording of a much higher signal-frequency span than could be achieved otherwise (without using excessive tape speeds), other problems had to be solved in achieving true video recording. One is the precise speed and tape head rotation for synchronization between recording and playback and for maintaining a playback speed that corresponds to that of the original picture.

In all types of magnetic-tape recordings the tape head characteristics alter the amplitude of the signal. Since the tape head contains a coil, it has a specific amount of inductance as well as a certain degree of reactance. An inductor's reactance is its opposition to the flow of an alternating-current signal. Unlike resistance, however, reactance is influenced by the frequency of the signal involved. This is evident by an inspection of the related equation:

$$\text{reactance } (X_L) = 6.28fL \tag{5-1}$$

where f = frequency, hertz
L = inductance of the coil, henrys

Thus, a coil having a value of 0.005 H has a reactance of 314,000 Ω at a frequency of 10 MHz. For the same coil at a frequency of 20 MHz, the reactance is doubled and is 628,000 Ω.

Consequently, higher-frequency signals developing across the inductor will have a higher amplitude than lower-frequency signals because of the reactance change.

Thus, there would not be a flat response in volume for all the signals encountered. In most high-fidelity audio tape recorders slight compensation is made for the uneven response, though in many instances it may not even be noticeable to the average ear. For video recording, however, where the frequency can range to 4 MHz, the reactance would be so much higher because the highest video signal frequency is 200 times greater than the highest audio signal frequency range. Hence, correction would be difficult to achieve. To avoid this problem encountered with an amplitude-modulated signal such as video, the AM video signal is converted to frequency modulation when recorded, and the FM is demodulated upon playback. (See also Fig. 5-1 and related discussions.)

With FM any increases in signal amplitude can be eliminated by signal limiters or clippers without altering the original signals (now FM) or without encountering the design problems in recording AM-type video signals.

As indicated in Appendix C, the color signal is made up of a luminance signal plus a chrominance signal. The luminance signal compares to the black-and-white portion of a televised scene and also contributes to the sharpness of picture detail and the brightness. It is this portion of the color video signal that is frequency-modulated prior to recording. The luminance signal, though only a part of the entire color picture signal structure, is the only signal that is converted to frequency modulation prior to recording on VCR tape. The remaining color signals are treated differently, as covered more fully in Sec. 5-7.

The luminance signal is also referred to as the Y signal. The structure of the Y signal consists of predetermined levels of each of the primary colors of red, blue, and green (the additive color system). Thus, the Y signal contains 0.59 green, 0.30 red, and 0.11 blue. These proportions are a close approximation of the color proportions to which the average eye responds.

5-7. DOWN-CONVERTED CHROMINANCE

In addition to the brightness (luminance) signal, a chrominance signal is also transmitted that contains the color information. The blue color signal is modified at the transmitter by subtracting from it the brightness (Y) signal. The resultant then becomes the $B - Y$ signal. A similar modification is applied to the red signal wherein the brightness signal is subtracted to form a signal known as the $R - Y$. Finally the green color signal is also altered and becomes $G - Y$. To maintain compatibility between color receivers and black and white, it is necessary to reduce the number of color signals before modulating the color carrier. This is done by combining the individual color signals to form two signals known as I and Q, with I representing *in phase* and Q the *quadrature* or *out of phase*.

To minimize interference between the signals during reception, the *I* signal is transmitted with one sideband extending to 1.48 (less than 2-dB decrease at 1.3) with a vestigial sideband approximately 500 kHz wide, as shown in Fig. 5-6. The *Q* signal is transmitted as a double-sideband signal with each sideband 500 kHz distant from the color subcarrier. The latter is suppressed before transmission and must be replaced in the receiver. The chrominance sidebands have a bandpass, as shown in Fig. 5-6.

The various hues of color appearing on the screen are produced by shifting the phase of the subcarrier generated in the receiver with reference to the burst signal of 3.58 MHz (actually 3.579545 MHz). In recording the chrominance signal on tape, the original method used in video tape recorders in television stations was to provide a guardband. The latter provided sufficient separation of the video signals to minimize interference. In modern VCR home systems, however, a different process is utilized wherein the chrominance signal is converted to a signal having a lower frequency. Thus, in the VHS system the 3.58-MHz subcarrier signal is heterodyned with a generated signal to produce a new signal that has a frequency of 629 kHz. (In the Beta machines the signal is down-converted to 688 kHz.)

When the luminance signal is converted to frequency modulation, the average deviation is approximately 3.4 to 4.4 MHz. Hence, the total bandwidth is 1 MHz, and the middle or resting frequency is 3.9 MHz. Both the FM luminance signal and the chrominance signal are then recorded simultaneously on the VCR tape. Because of the wide span frequency difference between 3.9 MHz and 629 kHz, there is a minimum of cross interference. During playback the luminance signal is reconverted to its original form, and the chrominance signal is heterodyned again to increase its frequency to the original before it was down-converted.

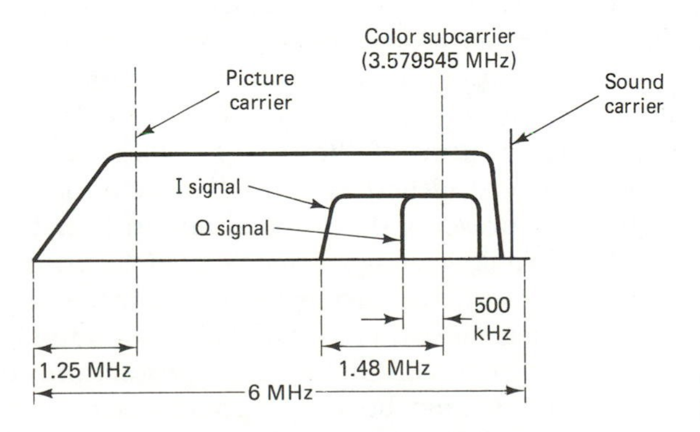

Figure 5-6 Bandpass of color signals.

5-8. BASIC VHS CIRCUIT INTERCONNECTIONS

A diagram illustrating the basic circuit interconnections for the VHS system is illustrated in Fig. 5-7. Here the individual blocks may represent several circuit linkages that relate to the same general function. Note that a single antenna input is present which receives both the VHF and UHF television station signals. A video signal selector switch channels the signals from the antenna directly to the VCR or to the TV receiver antenna terminals. As mentioned earlier in Chapter 1, the input system provides several options: (1) Station selection can be made at the television receiver whether or not the VCR is recording. (2) The switch can select the VCR as the tuner for either the television receiver or for recording purposes. (3) The VCR can tune to a particular channel and record it while another channel is viewed on the TV receiver.

The incoming signal from the antenna is applied to a VHF-UHF tuner in the VCR. As in a standard television receiver, the tuner contains a signal generator and by a heterodyne process forms intermediate-frequency signals. A circuit linkage between the IF stages and the tuner provides for an automatic frequency-control function. The demodulated audio and video signals are applied to output posts on the VCR. (See Fig. 2-6.) The RF signal amplifiers for record and playback modes plus the servo system are fed the *Y* signal, which has been frequency-modulated and has a center signal frequency of 3.9 MHz. The latter signal is the carrier signal without modulation. During modulation the carrier shifts above and below this frequency, as discussed earlier in Sec. 5-6.

The down-converted frequency signal of 629 kHz is also fed to the RF amplifier systems, and the luminance and chrominance signals are both amplified to the degree required for application to the recording/playback heads. The timing pulse is also applied to the RF amplifier/servo circuits and fed to the control head, as discussed earlier in Sec. 5-4. The audio head (either mono or stereo) records and plays back the sound signals from the upper 1-mm track. As discussed earlier in Sec. 5-3, high-fidelity techniques can also be employed for superimposing the audio on the video section of the tape for better quality. The bias for the erase head is also furnished by the RF system.

The SCC block represents the manual and automatic control circuits. A special tape slack sensing device is used to assure a smoother tape transport and to avoid the problems associated with a loose tape problem. With too much slack there is the danger of tape breakage, and the machine may shut off completely for protection. It is for this reason that a VCR tape should be stored to minimize spool movement within the tape housing. Before playing the tape, it should be inspected and the spools turned so there is no slack in the tape.

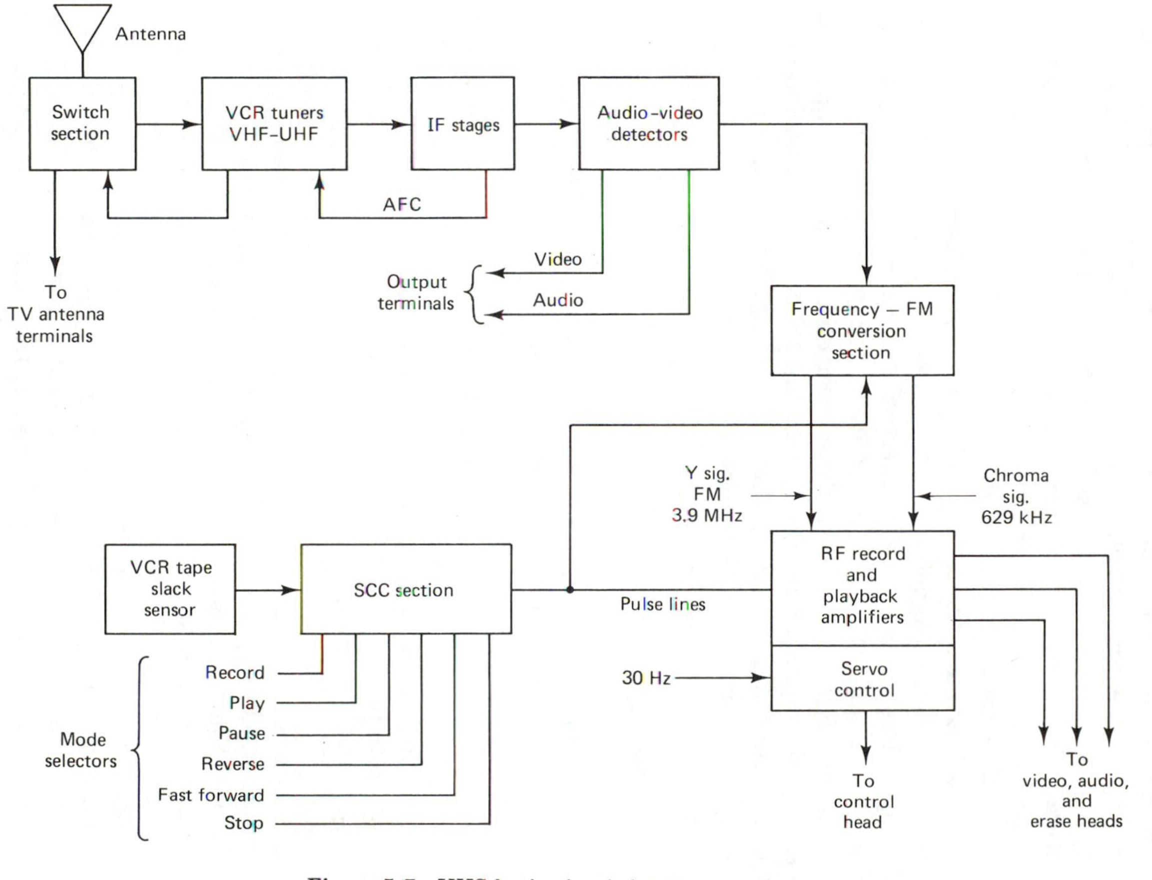

Figure 5-7 VHS basic circuit interconnections.

The SCC control circuits also sense the need for automatic overriding manual controls. Thus, the VCR can stop the machine at the end of a rewind period, for instance, or the operator can stop tape movement by push-button control. With solenoids tripping the control functions, the electronic signals are used to perform the necessary functions, particularly in the case of cordless remote control operation or during the period when the timer turns the machine on to record an event previously programmed into the VCR.

In the playback mode the recording heads become the playback heads and sense the magnetic impulses previously recorded on the tape. The signals thus obtained are transferred to the appropriate RF circuits where the luminance signals are converted from FM to AM and the chrominance signals are heterodyned and the frequency converted upward to its original value. The system control circuits also include the dc power supply and voltage divider sections that maintain proper potentials in all the circuits involved in the VCR.

5-9. VHS AZIMUTH AND PHASE

As discussed in Sec. 5-7, the chroma signal is down-converted, and the luminance signal is frequency-modulated. The design is such that there is a reduction of interference between the two (cross talk), even though no guardband separation is used. This video recording process is also termed *high-density* recording. Other characteristics of the VHS recording process include azimuth recording as well as phase inversion to provide for optimum recording with an absolute minimum of cross talk between both the luminance and chrominance signals.

Before the advent of the VCR, azimuth of recording and playback heads was already an important factor in audio tape recording. As shown in Fig. 5-8(a), the audio recording head is perpendicular to the tape travel, thus magnetizing true vertical recording areas. Thus, with the head gap perpendicular to the tape travel (90°) a horizontal tape travel would have the head gap in a true vertical position. With no tilt of the head gap to either left or right, the azimuth is considered to be at 0° azimuth. For a tape head not at 0° azimuth as shown in Fig. 5-8(b), the magnetized areas no longer are vertical but at a slant. Thus, there would be an azimuth difference between the head that recorded the tape in Fig. 5-8(a) and that in Fig. 5-8(b). If the tape shown in Fig. 5-8(b) were played back using the same head that recorded the material, the corresponding alike azimuth positions would permit the maximum audio range to be recorded and played back. If, however, the tape in Fig. 5-8(b) were played back with the head in Fig. 5-8(a), the high-frequency response of the reproduced music would suffer.

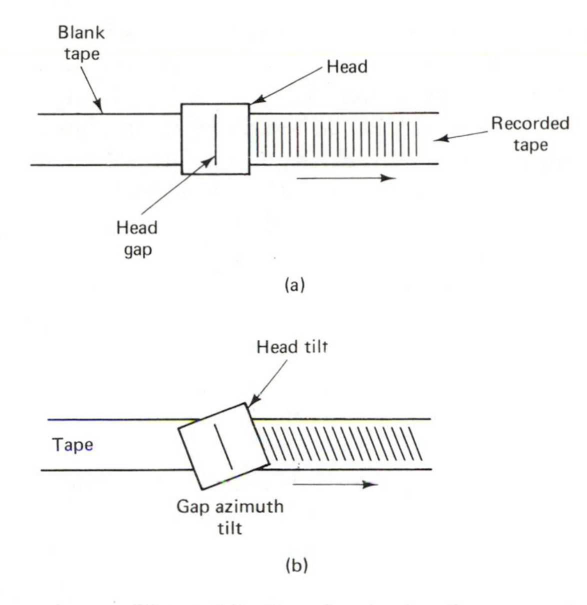

Figure 5-8 Tape head azimuth.

Azimuth recording factors were taken into consideration during the design of the VHS system. The azimuth recording for VHS utilizes an azimuth difference of ±6° for each video head on the drum. The result is a 12° azimuth difference between the first and the second primary video heads of the VCR.

Signal phase is also an important factor in video recording techniques related to the down conversion of the chrominance signal. As discussed in Sec. 5-7, the chroma signal was down-converted to a frequency of 629 kHz for the VHS system. The 629 kHz is derived by heterodyning (mixing) the 3,579,545-Hz chroma subcarrier signal with a 4,208,545-Hz signal. The 4,208,545-Hz reference signal used in the mixing process is phase-inverted and synchronized with the horizontal sync frequency locked in by the incoming horizontal synchronizing pulses of 15,734.264 Hz for color. The 15,750-Hz signal is used during black-and-white transmission. The down-converted signal of 629 kHz is thus 39.976 times the horizontal sync signal of 15,734.264.

During recording the phase inversion process involves advancing the phase of the signal on the first head by increments of 90° for each horizontal line trace. Thus, the phase changes for four lines, and at the fifth line the signal has again attained its original phase. The phase succession for the second head would then be 0°, 270°, 180°, and 90° and then 0° for the next horizontal line. For the first tape head, the sequence would be 0°, 90°, 180°, and 270°, after which the next horizontal line is again at 0° azimuth reference.

During playback the phase inversion process is reversed. Thus, the 4.208545-MHz signal used in heterodyning is again inverted in phase and the mixing process repeated using the 629-kHz signal. The process now restores the original 3.579545-MHz chroma signal. Thus, during the playback mode, the 629-kHz signal and the 4.208545-MHz signal are shifted in phase in sequence again to recover the 3.579545-MHz chroma signal in its original phase. During successive line traces the phase difference between the two heads mandates that the recording made by one head is not played back by the other head. Consequently, cross talk between the two heads during playback is minimized.

5-10. VHS SECTIONS AND TAPE LOADING

Modern video cassette recorders are much more compact and are less heavy than the first ones that appeared on the market. There is also a difference in the specific locations of circuitry among manufacturers. Exact identification of sections, circuits, and component parts can be obtained by reference to the schematic of the particular VCR in question. Generally, however, the sections can be readily identified by the type and grouping of components, as detailed later in Sec. 5-11. For the VHS system a larger area must be assigned for the cassette than for the Beta. Otherwise, the circuit and systems layout may be quite similar. The essential differences between the two systems are the tape speed, the down-convergence signal frequencies, azimuth, and phasing, as discussed earlier in this chapter. In addition there is also a difference in the tape loading, as discussed in this section.

A representation of sections in a VCR is shown in Fig. 5-9. This is typical of the left-front loading system. The clock and timer display is sometimes at the center of the front panel and often at the upper right. Thus, the circuitry and timer setting buttons are also at the right, as shown in Fig. 5-9. With cable-ready VCR units the manual and remote control channel selector section may be in the general area shown. This section permits the selection of a specific channel sequence to accommodate the channels available from the local cable company. For those VCRs with only a limited station selection capability, the fine-tuning set screws and station selection tuning would be present, as shown later in Sec. 5-11.

The next section contains the IF stages, video and audio demodulators, the FM modulators, and the down converters for the color signals as well as the luminance and chroma signal processors. In some VCRs this circuitry may be divided into several individual sections. The power transformer is located at the right rear, and this unit feeds the power supply section. The servo control is at the left rear, and all these sections

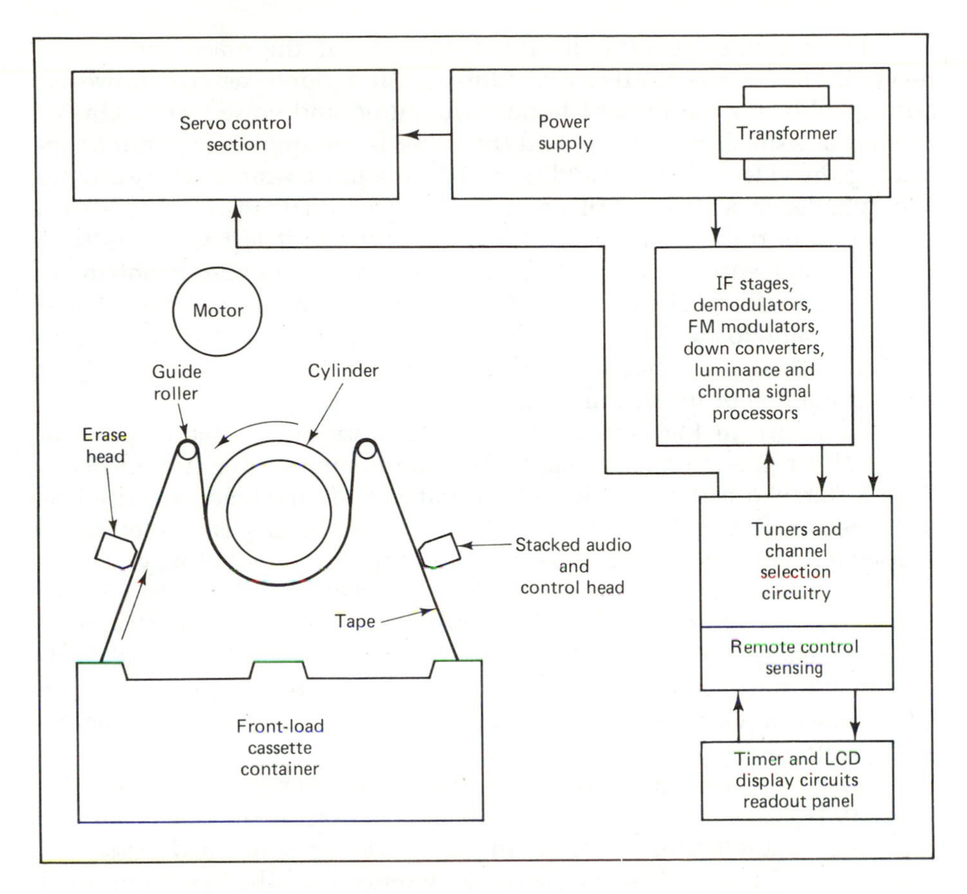

Figure 5-9 Typical chassis layout and tape path (VHS).

are interconnected as shown by the lines and arrows. In the VCR the interconnections would consist of multiwire cables and plug-in sections so that any particular section can be unplugged and removed for convenience in servicing.

The cylinder that contains the video heads for recording/playback is located at the left, above the cassette holder, as shown in Fig. 5-9. The motor for spinning the cylinder is shown at the upper left. This motor is linked to the cylinder by belt drive. Another motor (usually next to the front-load cassette retainer enclosure) turns the wheels that draw the cassette sufficiently far into the machine to engage the loading mechanism. The latter loads the tape onto the cylinder, as explained later. This same motor, when used, also ejects the cassette when this operational mode is initiated by pushing the eject button. The tape head cylinder is surrounded by tape guideposts, control and erase heads, as well as audio heads.

In an audio cassette the fixed position of the erase and record heads simplifies the loading procedure. With a video cassette, however, additional heads are present besides the audio and erase heads. One of the latter is the control head, and the other is the spinning cylinder containing the video heads. In addition, the tape must wrap halfway around the cylinder in a slanting tilt, which would be difficult to achieve with simple guide rollers. An abrupt tilt of the tape as it left the cassette at one end and entered at the other would bind and cause problems in proper tape feed. Hence, the VCRs are designed so that the tape is automatically loaded onto the cylinder and around the various guide rollers gradually so that the tilt is formed without causing problems such as binding or uneven movement.

As shown in Fig. 5-9, the VHS system uses an M-loading pattern. The latter refers to the M-shape the tape assumes after it is in place. This differs from the Beta where a U-pattern of tape loading is used, as discussed in Sec. 6-10. The M-pattern of tape loading is formed automatically, with the guide rollers positioning the tape halfway around the cylinder and at the proper slant. The loaded tape shown in Fig. 5-9 illustrates the basic principle involved, though in the actual system other rollers or tension levers may be used. Since each VCR model differs in the number of guide rollers, positioning posts, and so on, the VCR under test or checking should be inspected for the particular design features employed.

When the record or playback button is depressed, the cylinder spins at 1800 rpm, as discussed earlier in Sec. 5-2. At the same time the tape movement originates from the left cassette spool and passes an erase head. The tape then engages the cylinder with the heads mounted on it, and the tape travels in a slanting position, as discussed earlier. Next the tape passes the audio and control heads and then enters the take-up reel of the cassette at the right.

During the fast-forward and rewind modes the tape in the VHS system is disengaged from the cylinder. During the review-cue mode, however, the tape configuration around the cylinder is maintained.

5-11. IDENTIFYING SECTIONS

The most convenient method for identifying the sections in a VCR is to acquire a schematic of the particular unit in question. A schematic (obtained from an authorized dealer or a wholesale house or directly from the manufacturer) lists and identifies component parts and specific sections involved. In lieu of a schematic, identification of a particular section is facilitated by a knowledge of the type of components usually found in that section. Another clue is to ascertain what sections are connected to the input and the output of the specific section to be identified.

The two sections most easily identified are the tuner and power supply. The tuner is identified as that section to which the input television antenna or cable input is connected. The tuner section also may contain a series of set screws for station selection, as shown in Fig. 5-10. The tuner also connects to the IF stages, and thus the latter can be localized by checking tuner output lines.

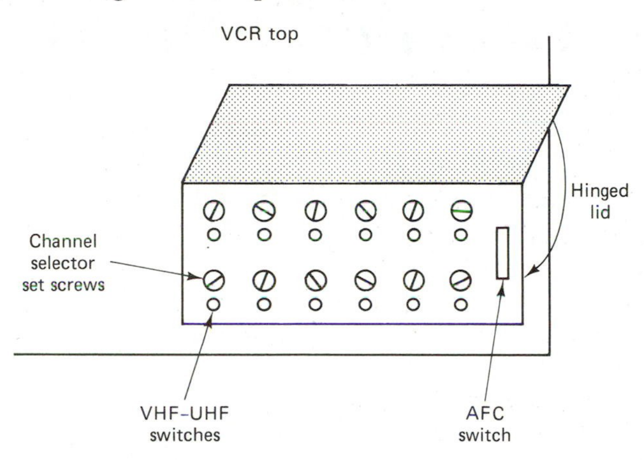

Figure 5-10 Channel selector set screws.

The power supply is localized by noting the route of the input ac line and plug inside the chassis. This ac line also goes to the on-off switch and the power transformer. The output lines of the transformer connect to the rectifiers and the filter capacitors of the power supply and help pinpoint this area.

The servo control section is fed by the power supply, and part of the servo system feeds the control sections of the system and links to the control head. The video recording heads are connected to the up-conversion and FM demodulators during the play mode; hence, these circuits can be traced back from the heads and identified. Similarly, during recording, the heads are fed by the output circuits of the down converters and FM modulation circuits.

Other sections are similarly traced. The leads from the reverse, fast-forward, pause, stop, play, and record buttons can be followed to the respective relays, circuit sections, and switch-trip levers. Some of the latter are tripped for automatic rewind, timer control for program recording, and other automatic functions. Here, again, the related circuitry and wiring can be traced to circuit sections that process the signals for such control purposes.

6

The Beta Recording System

6-1. INTRODUCTION

To better understand the methods utilized for recording video and audio in the Beta video cassette recording system, reference should be made to Sec. 5-1. (tape recording basics) of Chapter 5. In that section the fundamental principles of magnetic recording as applied in cassette tapes are briefly outlined. That discussion provides an essential foundation for the comprehension of the more involved aspects of video recording. Although the VHS recording principles differ to a considerable extent from the Beta type, the basic procedure of implanting magnetized segments on a moving tape are basically the same. Thus, an overview of cassette recording principles aids considerably in understanding video recording and playback factors. Although the BETA and VHS systems are not compatible, there are a number of similar aspects in the basic system. Hence, some of the data in this chapter parallel several of the discussions found in Chapter 5.

6-2. BETA TAPE SCAN SYSTEM

As discussed in Sec. 5-2, the tape heads in audio cassette systems have a narrow vertical gap in their center area that faces the tape. The tape head gap must be as narrow as possible to obtain the widest frequency range in recording. For cassettes, the audio signals may range to 15 or 20 kHz, and good fidelity is achieved with a cassette tape speed of 4.5

cm/sec. Of importance also in high-fidelity recording is the quality of the tape used, and a considerable improvement in wide-band response (particularly in the higher-frequency signals) was achieved with the advent of chromium-dioxide magnetic tape material. Subsequently, other magnetic material compounds were developed for improved audio response, and advantage was taken of the developments in the manufacture of VCR tapes.

Since the frequency range of video signals extends to 4 MHz, a VCR must be capable of handling signals within that span. Since the latter has an upper limit some 200 times that of good audio signals, an increase in tape speed and use of a more narrow head gap are not sufficient to record and play back from such a range. Consequently other measures must be taken to enable the VCR to handle the video signals without undue loss in the upper regions that extend to 4 MHz. For the Beta system the tape head gap is 0.6 μm compared to 0.3 μm gap for the VHS system. To obtain the necessary signal bandpass, the method of tape scan was changed radically from the cassette process wherein the tapes slip past a stationary record/read head. As with the VHS system discussed in Chapter 5, the Beta system of scan uses a rapidly spinning tape head, the basic principle of which is shown in Fig. 6-1.

As shown in Fig. 6-1(a), a rotating drum is used that spins in the direction indicated. As with the VHS system the tape wraps around the drum in a Ω configuration. The wrap-around aspects are more clearly indicated in Fig. 6-1(b). Note that the tape wraps around the drum in a diagonal pattern held in place by a guide at each side of the drum. The drum uses a minimum of two tape heads placed directly opposite each other. The drum rotates at 1800 rpm, the same speed as in the VHS system. Consequently, the combination of the rotating drum and tape movement (2 cm/sec) produces a relative speed of 6.9 m/sec (69 cm/sec). With these speeds and design the Beta system thus records and plays back the complete frequency range of the video signals. (Here it is assumed that the standard speed of B II is used rather than the B III.)

Despite the similarities between the Beta and VHS system, there are sufficient design differences to make them incompatible. In the VHS system a cylinder is used instead of a drum, though both spin at 1800 rpm. As mentioned earlier, there is a gap difference between the two systems, though the actual tape speed is the same for each. In both systems the tape speed can be changed to obtain a longer recording time. With the latter, however, picture and sound quality suffer. (See Sec. 6-5.)

Several additional heads can be used to improve auxiliary functions such as freeze frame or reverse scan. When two heads are employed, each records or plays back one television field containing 262.5 lines. (See Sec. 1-2, Fig. 1-7.) Because each head records or plays back

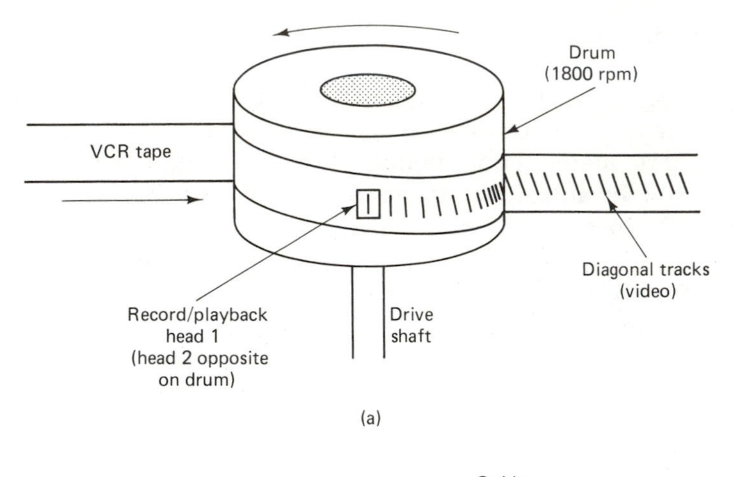

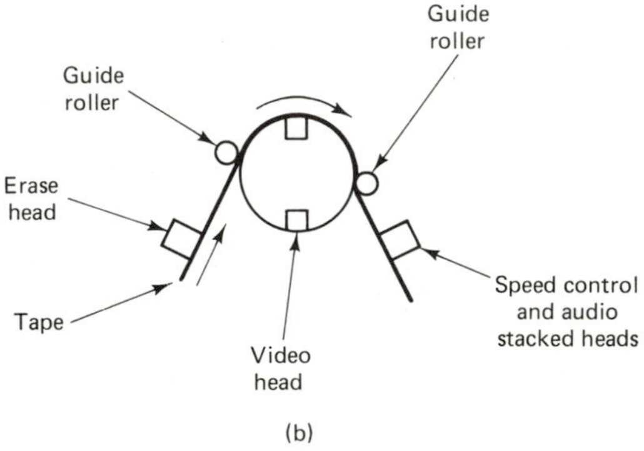

Figure 6-1 Beta scan system.

one television field, the combined heads generate 525 lines (a complete video frame).

The diagonal (or slant) recording is known as the *helical scan system*, and the tape guides must hold the tape at the proper slant across the drum. As shown in Fig. 6-1(b), the erase head is in a fixed position, and its sole function is to clear the tape of any previous recording before the tape reaches the recording drum. The erase head is inoperative during playback. An additional head is also present for audio recording and pickup. In the basic system the audio is in the upper portion of the tape, as discussed in Sec. 6-3. Another head is also used for tape speed control, and this utilizes the lower portion of the track. The control head and the audio head can be stacked to occupy less space. During playback the control head senses the prerecorded timing pulses, and these are used to regulate the speed in the manner more fully discussed later in this chapter. As with the erase head, the control and audio heads are stationary.

6-3. BETA SOUND FIDELITY

Modern VCR units are capable of producing high-fidelity sound quality when the sound output is channeled to a high-fidelity sound system containing high-quality amplifiers and speakers. In earlier VCR systems, however, a mono sound track, as shown in Fig. 6-2(a), was the only sound source. Since the audio head is stationary and the tape speed is slow (2 cm/sec as compared to 4.5 cm/sec for cassettes), the audio sound quality suffers. Thus, if there is some background noise as well as wow and flutter present in older VCR models, the poor quality may be caused by the inherent limitations present. The wow and flutter can be minimized by recording and playback at the standard speed. Of some help, also, are the usual measures of cleaning the heads and oiling the components of moving portions such as guide wheels, rollers, as well as other measures discussed in various sections of this text.

The audio strip in the upper section of the tape is only a small part of the total tape width. The VCR tape is 12.7 mm wide (½ in.), and only a narrow 1-mm strip along the top is allocated to the sound recording. This section can be subdivided into three distinct sections for

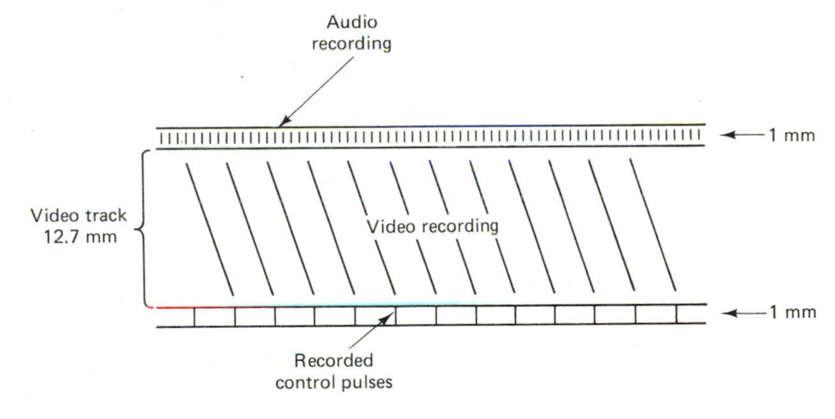

(a)

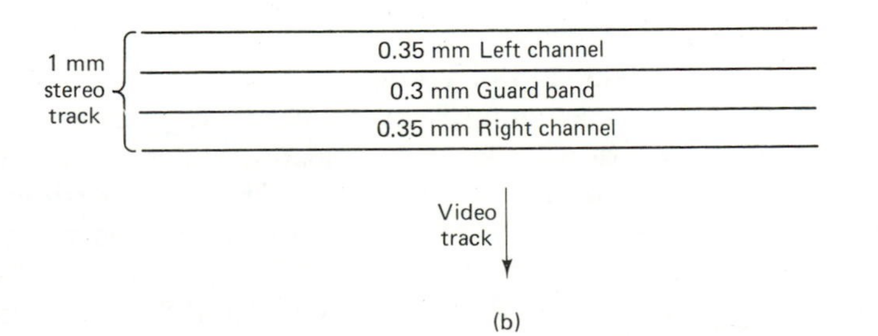

(b)

Figure 6-2 VCR tape recorded segments.

stereo, as shown in Fig. 6-2(b). Here two 0.35-mm sections are utilized for the right and left stereo channel, and a center 0.3-mm space is placed between the two channel sections to minimize cross interference. (see Sec. 5-3.) With such stereo, however, the actual sound quality was not improved, though the stereo effect was more desirable than mono. Again, the best results were obtained when the highest recording speed was used.

As mentioned earlier in Sec. 5-3, stereo systems were improved considerably with the advent of recording the sound on the video track as well as on the 1-mm sound track initially used. In the initial VHS system two additional heads were used in the cylinder to superimpose the audio on the same track as the video. Similarly, Beta employs a system that uses the video portion of the tape to record video and audio simultaneously. Each of the two heads acts as a video/audio recording and playback head in a video carrier system of the audio signals. Thus, the audio signals also benefit from the advantages of the increased relative speed attained by combining the rotating drum with the lateral tape movement. The original upper audio band is retained so that the playback of older tapes will still be possible. Thus, the high-fidelity advantages obtained by placement of both the video and sound on the video track are procured for newly made tapes while still permitting the playback of older tapes. For the older tapes, however, the high-fidelity features are lost because of the inability of the older machines to demodulate the video signals in order to obtain the high-fidelity sound. (See Sec. 12-2.)

6-4. BETA SERVO CONTROL

In any video tape recording system precise speed control of the rotating head and tape movement must be present. Without such control the playback of any recorded material would be out of synchronization and would be unsatisfactory. As mentioned in Sec. 5-4 for the VHS system, lack of synchronization to even a slight degree results in poor picture quality as well as line streaking interference patterns.

The type of control uses one of several servo systems involving a comparison of drum rotation to synchronization pulses recorded on a control track at the bottom of the VCR tape. The servo system compares the drum rotation with the recorded control pulses for timing coincidence. If there is any deviation between the two, speed correction is immediately made. Several methods can be utilized for servo control including the servo control of drum rotational speed or the use of the capstan drive roller to alter tape speed when correction is necessary during playback.

As mentioned earlier, when two rotating heads are present on the drum, each head scans one field of the television picture frame in $\frac{1}{60}$ sec. Thus, each head traces out one of the two fields making up a frame of 525 scan lines (262.5 lines per field). Since the drum rotates at 1800 rpm, the rotation in 1 sec is 1800/60, which equals 30 revolutions per second. The designation often is 30 Hz, though strictly speaking hertz denotes *cycles per second* rather than *pulses per second* or *revolutions per second*.

The control track pulses are formed and precisely timed by the vertical sync pulses present in the incoming television signals. Essentially, these pulses have a rate of 60 Hz, the same as the 60-Hz power mains, though a slight difference exists between the sync signals for color compared to the black-and-white signals. (See Appendix C.) The 60-Hz sync pulse train is applied to an electronic circuit (a divider) that halves the 60-Hz frequency and thus produces a 30-Hz pulse train. The latter is also termed the control (CTL) signal. It is this CTL signal that is recorded on the control track illustrated in Fig. 6-2. The audio and control heads are stacked as shown in Fig. 6-3.

As shown in Fig. 6-4, the basic principles of a typical servo system utilizes a compare and correct sequence. The 30-Hz signals are recorded on the control track. The rotational speed of the drum is sensed by a small magnet segment embedded in the drum, as shown in Fig. 6-4. A coil is mounted so that it passes under the upper magnet. Consequently, a magnetic pulse is created each time the magnet passes over the coil. Hence, the latter senses the rotational speed and transfers this information to a comparison circuit. This section evaluates any difference between the 30-Hz rotation of the drum and the 30-Hz pulse train impressed on the control track during recording. For any variation that occurs during playback, the drum rotational speed is corrected by the speed control section using a magnetic brake assembly. The speed is thus increased or decreased sufficiently to achieve perfect coincidence with the control track frequency.

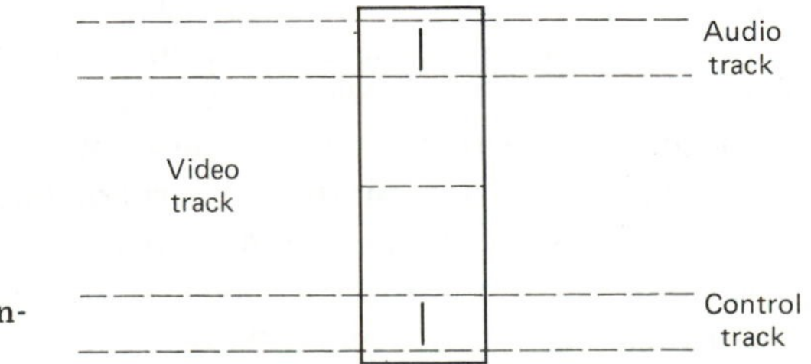

Figure 6-3 Stacked audio and control heads.

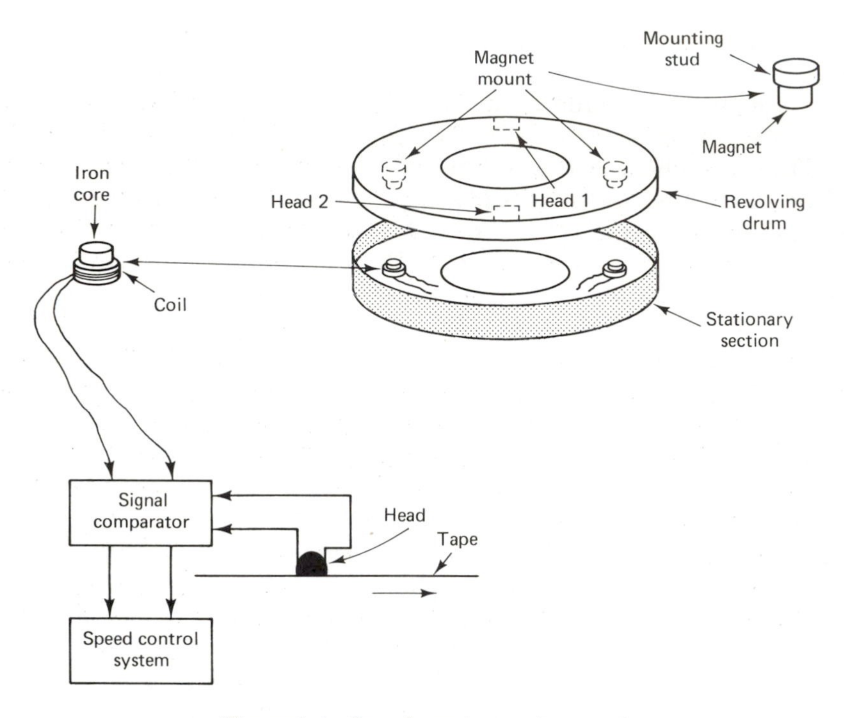

Figure 6-4 Speed sensing and control.

6-5. TRACKING CONTROL FACTORS

One essential item in a VCR is a tracking control for obtaining proper playback of tapes that were recorded on another VCR. A typical location of such control is shown in Fig. 6-5 where other front-panel switches and knobs are shown. In some VCR machines the tracking control may be a conventional knob rather than a thumb wheel. For the control shown, the tracking control wheel is at the bottom of the front panel toward the right. Other models of a VCR may have the tracking control at a different position. (See also Fig. 5-5.) The other panel switches and knobs shown in Fig. 6-5 are typical of those present in most machines, but the exact placement in various models will vary considerably.

A tracking control is essential because of the slight differences that may exist between VCR units of different manufacturers, even though they also utilize the Beta principle. Thus, a commercial tape containing a motion picture or a recorded tape made by a neighbor who also has a Beta machine may present problems when played on your machine. This comes about because of slight timing differences between machines (sometimes even from the same manufacturer).

The tracking control utilizes the synchronization pulses recorded

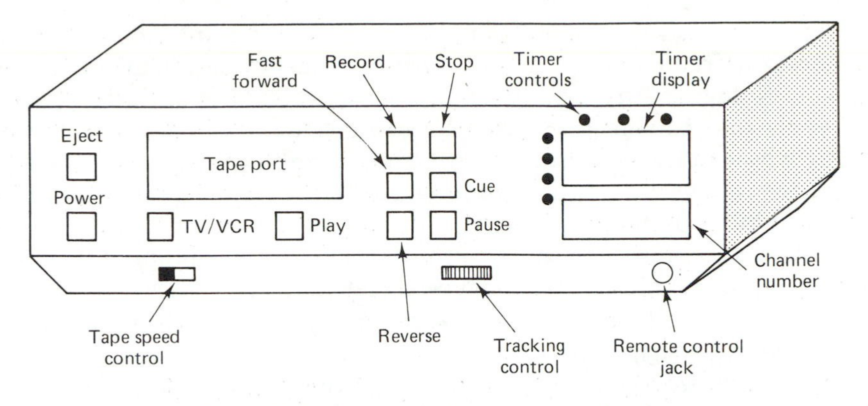

Figure 6-5 Front-panel switches and knobs.

on the control track to make a slight correction to achieve compatible control when playing a different tape. The primary purpose of this tracking control is to make such corrections during playback. The control usually has a marking that indicates the standard or central setting. After playback of a tape where correction was necessary, the tracking control should be repositioned to the standard setting.

Tracking control adjustment is necessary when a poor picture quality is present and when black streak lines appear across the screen during playback. The tracking control is most effective for tapes recorded at the standard speed of Beta II. The tracking control is less effective if the recorded material uses the slower Beta speed indicated as B III.

6-6. FREQUENCY-MODULATED LUMINANCE

Despite the advantages of a rotating drum with multiple recording heads plus the moving tape, additional problems had to be overcome to achieve good-quality video tape recording. When magnetic-tape recordings are made, the tape head with its coil and metal core has characteristics that change the amplitude of the applied signal for different frequencies. Each recording/playback head contains a coil having a specific amount of inductance and hence an opposition to signal flow (termed *reactance*). An inductor's reactance creates opposition to signals in a fashion similar to resistance, except the latter does not change for signal-frequency changes. [See Eq. (5-1) and related discussions in Sec. 5-6.]

A higher reactance produces a larger voltage drop across it, and this is the contributing factor to an uneven recording and playback mode. Since the highest video signal frequency is 200 times greater than the highest audio signal frequency, flat-response recording problems are increased. A satisfactory solution is to convert the amplitude-modulated signal to a frequency-modulated type. Frequency modulation has several advantages. Any unwanted increase in signal amplitude (such as noise signals) can be minimized by using signal limiters or clipper circuits. These only permit a specific signal amplitude to pass through. Since the signals are frequency-modulated the clipping process of amplitude modulation does not alter the desired signal information.

As indicated in Appendix C, the color signal is composed of both a luminance signal plus a chrominance signal. The luminance signal can be compared to the black-and-white segment of a scene televised in color. The luminance signal also relates to the brightness of the picture as well as sharp picture detail. It is this luminance signal that is frequency-modulated in a VCR prior to the recording process. During playback the FM process is reversed by demodulation, and the original type of signal is restored.

The luminance signal is also designated as the *Y* signal and consists of specific levels of each of the primary colors red, blue, and green (the additive color system). The *Y* signal contains 0.59 of green, 0.30 of red, and 0.11 of blue. These proportions are based on those color levels perceived by the average human eye.

6-7. HETERODYNED CHROMINANCE

In addition to the luminance (*Y*) signal a chrominance signal is also transmitted, and it is this signal that contains the color information for a television scene. The blue color signal is modified at the transmitter by subtracting from it the brightness (*Y*) signal. The resultant signal is termed the R - Y signal. Similarly, the red signal is modified by subtracting the brightness signal to form the R - Y signal. The green color signal is similarly altered and becomes the G - Y signal. In the interest of conserving spectrum space so the compatibility between color and black-and-white receivers would not be disturbed, the color signals were combined to form two signals known as the *I* and *Q*, with *I* representing *in phase* and *Q* denoting *quadrature* or out of phase.

To minimize interference between the various signals in the receiver, the *I* signal is transmitted with one sideband extending to 1.48 and with less than a 2-dB decrease at 1.3. A vestigial sideband of approximately 500 kHz is present, as shown earlier in Fig. 5-6. In transmission the two signals are sent as a double-sideband type with each sideband 500 kHz distant from the color subcarrier. The latter is suppressed before transmission of the other signals and hence must be generated within the receiver and combined with the chrominance signals. The chrominance sidebands have a bandpass as shown earlier in Fig. 5-6.

In color reception the various hues appearing on the screen are produced by changing the phase of the subcarrier generated in the receiver with reference to the burst signal of 3.58 MHz. When the chrominance signal is recorded on the VCR tape, it differs somewhat from the original method used in video tape recorders in television stations. In the latter a guardband on the tape separated the video signal that could cause interference. In VCR home systems, however, the process differs, and the chrominance signal is converted to a lower-frequency signal. Thus, the 3.58-MHz subcarrier signal is heterodyned with another signal generated to produce a signal having a frequency difference of 688 kHz for the Beta system (in the VHS machines the signal is down-converted to 629 kHz).

As mentioned earlier, the luminance signal was converted to frequency modulation. The average deviation for the FM signal is approximately 3.5 to 4.8 MHz. Hence, the total bandwidth is 1.3 MHz, and the middle or resting frequency is 4.15 MHz. Once the luminance signal has been changed to FM and the chrominance signal to a down-converted signal, both are recorded simultaneously on the VCR tape. Because of the wide span frequency difference between the 4.15-MHz signal and the 688-kHz signal, there is a minimum of cross interference during playback. For the latter process, the luminance signal is reconverted to its original form, and the chrominance signal is heterodyned again to bring its frequency back to the original frequency.

6-8. BASIC BETA CIRCUIT INTERCONNECTIONS

The basic circuit interconnections for the Beta system are shown in Fig. 6-6. Except for the differences in the frequencies of the down-conversion chrominance and the FM luminance, the sections are similar to those shown in Fig. 5-7. As with the latter, the individual blocks shown in Fig. 6-6 may represent several circuit linkages that related to the same general function. A single antenna input is provided on the VCR, and both the VHF and UHF channels are received here. A selector switch either routes the video signal from the antenna directly to the VCR or to the TV receiver antenna terminals. As discussed in Chapter 1, the input system of a VCR provides several options: (1) Station selection can be made with the TV receiver tuner whether or not the VCR is recording. (2) The switch can select the VCR as a tuner for either the television receiver or for recording purposes. (3) The VCR can tune to a particular channel and record it while another channel is viewed on the TV receiver.

The television station signals received from the antenna are applied to a VHF-UHF tuner in the VCR. The latter, as in a television receiver,

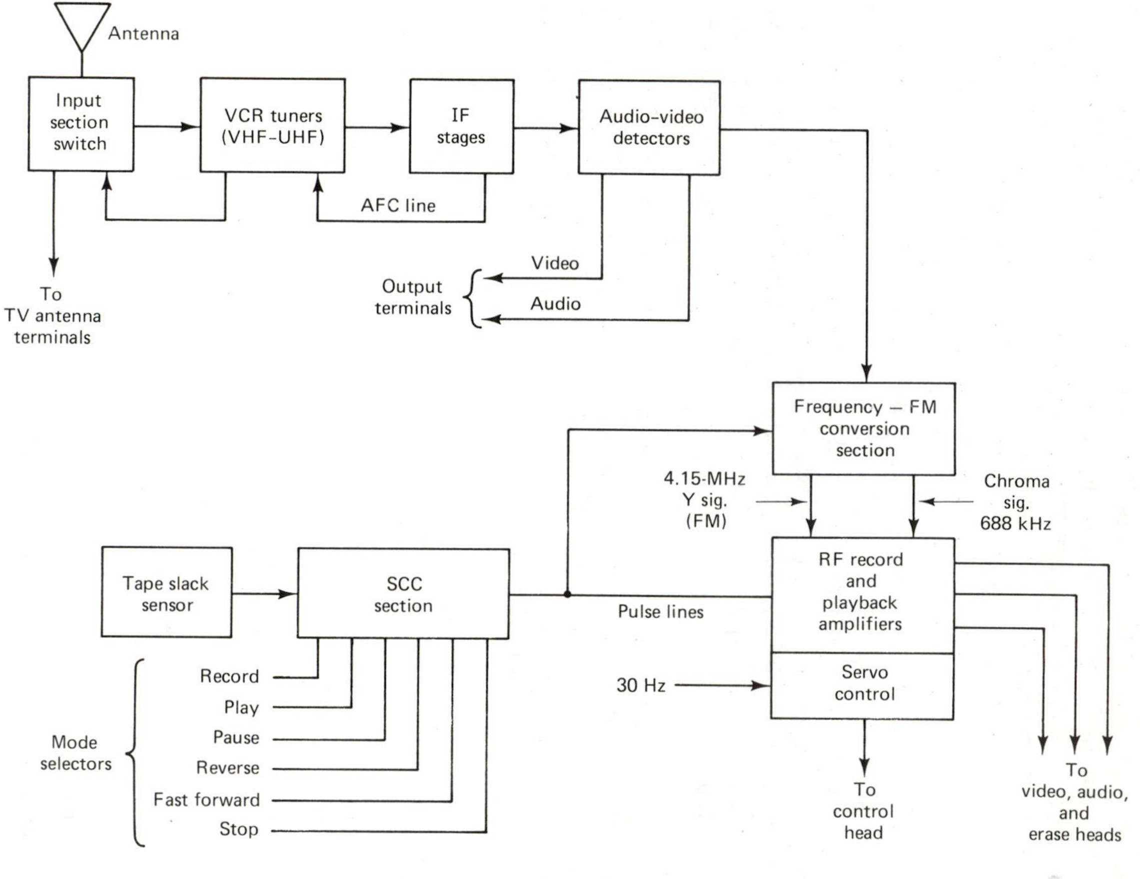

Figure 6-6 Beta basic circuit interconnections.

contains a signal generator that produces a signal for heterodyning purposes with the incoming signal. The resultant intermediate-frequency signal is then applied to IF amplifier stages. An automatic frequency control system is utilized in the IF stages, and a correction signal is applied to the output posts on the VCR, as shown earlier in Fig. 2-6. The RF signal amplifiers for record and playback functions as well as the servo systems receive the luminance signal that has a center signal frequency of 4.15 MHz. The latter signal is that of the FM carrier without modulation. During modulation the carrier shifts above and below this frequency, as more fully detailed earlier in Sec. 6-6.

The chrominance signal which has been down-converted to 688 kHz is also applied to the RF amplifier system, and both the luminance and chrominance signals are amplified as necessary for application to the recording/playback heads. The timing control pulses are also applied to the RF amplifier system and also fed to the control head, as discussed earlier in Sec. 6-5. The audio head (either mono or stereo) both records and plays back the sound signals from the upper 1-mm portion of the VCR tape track. As discussed earlier in Sec. 6-3, high-fidelity techniques can also be employed, including the recording of the sound signals on the same track used by the video signals. The RF system section also generates the bias for the erase head.

The SCC block represents the system's manual and automatic control circuits. One section is used to sense any VCR tape slack to minimize the problems associated with loosely threaded tapes in the tape container assembly. Excessive slack can cause tape breakage and tape-travel malfunctions. Some VCR units have sufficient built-in sensitivity to provide for complete VCR shutoff if there is any slack in the recording tape. Because of the problems that may arise with loose VCR tape, the cassettes should be stored in their original containers to minimize spool movement. Before playing the tape, it should be inspected, and if any slack appears, the spools should be turned to eliminate the slack.

In a VCR the function controls can be manually operated or automatically initiated within the VCR. Thus, the VCR circuitry can stop tape movement and shut off the machine drive at the end of a rewind. Similarly, the operator can stop tape movement by pushing the stop button. The VCR utilizes solenoids for initiating control functions and for permitting use of electronic signals to perform the necessary control-function sequences. VCR circuit control is particularly useful where cordless remote control systems are employed or during the period when the timer turns the machine on to record an event previously programmed into the VCR.

In the playback mode the recording heads become the playback heads and sense the magnetic impulses that have been previously recorded on the tape. As the playback heads read such signals, the latter are transferred to appropriate circuits for reconversion of the luminance signal from FM to AM. At the same time the chrominance signal is heterodyned with a signal generator within the VCR, and the chrominance signal frequency is reconverted upward in frequency to its original value. The system control circuits also include the dc power supply and voltage-divider sections that maintain the proper potentials in all the VCR's circuits.

6-9. BETA AZIMUTH AND PHASE

As mentioned earlier in Sec. 6-7, the reduction of interference between the luminance and chroma signals is reduced considerably by down-converting the chroma signal and changing the luminance signal to frequency modulation. Since the two signals are recorded adjacent to each other (without a guardband between them), this recording process is termed *high-density* recording. To hold cross talk at a minimum, however, two other procedures are employed during recording. One of these is azimuth recording, the same as utilized with the VHS system described in Sec. 5-9. The other procedure involves phase conversion, as described later in this section.

From the early days of tape recording audio signals, common procedure was to position the tape head so that the elongated magnetic gap was perpendicular to the tape travel. This factor was shown earlier in Fig. 5-8(a). Thus, if the tape travels in a horizontal plane, the head gap will be in a true vertical position without any tilt to either the left or right. Such a true vertical position is considered to be a 0° azimuth. The latter is illustrated in Fig. 6-7(a), and a tape head not at 0° azimuth is shown in Fig. 6-7(b). If the tape shown in Fig. 6-7(b) were now played back with the same head that recorded it, the identical azimuth relationships between the head gaps and the recordings would coincide and would produce the maximum audio range possible with the equipment used. However, if the tape recorded by the head in Fig. 6-7(b) were played with the head illustrated in Fig. 6-7(a), the results would be inferior. With a slight azimuth displacement, the 0° azimuth playback head would reproduce sound with attenuation of high-frequency signals. With a sufficiently significant azimuth displacement, no playback is achieved with a 0° azimuth tape head.

Azimuth-change recording is involved in both the Beta and VHS systems. For the Beta system azimuth recording involves an azimuth difference of ±7° for each main video head on the drum. The result is a 14° azimuth difference between the first and second main video heads of the VCR.

Another important process in video recording techniques relates to

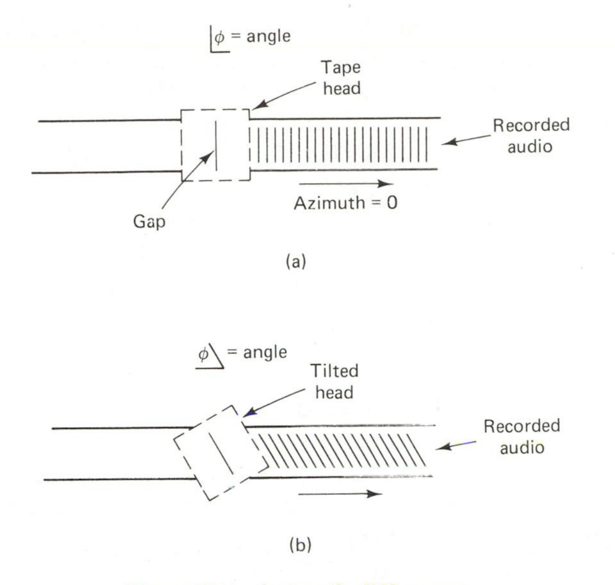

Figure 6-7 Azimuth differences.

changing the phase of specific signals. As discussed in Sec. 6-7, the chroma signal is down-converted to a frequency of 688 kHz in the Beta system. The 688-kHz frequency is obtained by a circuit mixing process called *heterodyning*. Thus, the 3.579545 MHz chroma subcarrier signal is heterodyned with a 4,767,545-Hz signal. This 4,767,545-Hz signal (also termed a *reference* signal) is inverted in phase and synchronized with the horizontal line frequency. The latter is locked in by using the incoming horizontal synchronizing pulses having a timing rate of 15,734.264 Hz for color. The down-converted 688-kHz signal is thus 43.7 times the horizontal sync signal of 15,734.264 Hz. In a black-and-white receiver the horizontal sync signal has a frequency of 15,750 Hz. This frequency is close enough to that used for color to be processed without difficulties by the synchronizing circuits. Thus, color sets can receive black-and-white pictures in true compatibility. Similarly, the black-and-white sets can get the color signals but receive them in black and white.

Phase changes occur during recording because the phase of the signal on the first primary video recording head is advanced by increments of 90° for each horizontal line trace. Thus, the phase changes for four lines, and at the fifth line the signal phase is the same as the initial one. Hence, the first head has a phase sequence of 0°, 90°, 180°, and 270° after which the next horizontal line again has a 0° phase reference.

The phase-change sequence for the second head is 0°, 270°, 180°, and 90° and 0° for the fifth horizontal line. During playback the phase inversion sequence is reversed. Thus, the 4.27-MHz signal that is used in the mixing process is again inverted in phase, and the heterodyning process is repeated using the 688-kHz signal. This process results in the restoration of the original 3.579545-MHz chroma signal. Thus, during the playback mode, the 688-kHz signal and the 4.27-MHz signal are shifted in phase in sequence again to recover the 3.579545 chroma signal in its original phase. Because of the phase difference between the two heads for a successive line trace, the recording made by one head is not played back by the other head, and consequently cross talk between the two heads in minimized.

6-10. BETA SECTION IDENTIFICATION

As the state of the VCR art has progressed, the units have become more compact. Some models are sufficiently lightweight to be portable for use with cameras. For many types, identification of the various sections can be made by reference to the schematic for the VCR. With table models, however, the sections can be identified by inspecting their general appearance and noting the type of components utilized, as detailed earlier in Sec. 5-11. For the Beta system a somewhat narrower cassette loading space is needed than for the VHS system. Generally, however, the placement of sections and the circuitry are similar. As mentioned earlier in this chapter, the primary differences between the VHS and Beta systems include the tape speed, the frequency of the down-converted signals, azimuth, phasing, and tape loading format.

A typical section layout for a Beta system is shown in Fig. 6-8. This is the general sectionalization for a left-front loading system. The clock and timer display may be at the center of the front panel or at the right, with the circuitry usually behind this area, as shown in Fig. 6-8. The tuner section is near the timer, and the set-screw channel selector section, shown earlier in Fig. 5-10, can also be present here. Once the stations have been selected, the stations can be selected by push buttons, one for tuning upward in channel numbers and the other downward. For complete cable-ready systems a selection can be made of a specific sequence of channels as furnished by the local cable company.

The next section contains the IF stages plus the video and audio demodulating system. Included in this section may be the FM modulator and the down-converter circuits for processing the color signals for recording as discussed earlier. The power transformer is located at the back near the power supply, and the servo control is to the left, as shown. All sections are interconnected by flexible multiwire cables with plugs for disconnecting any section as needed for servicing.

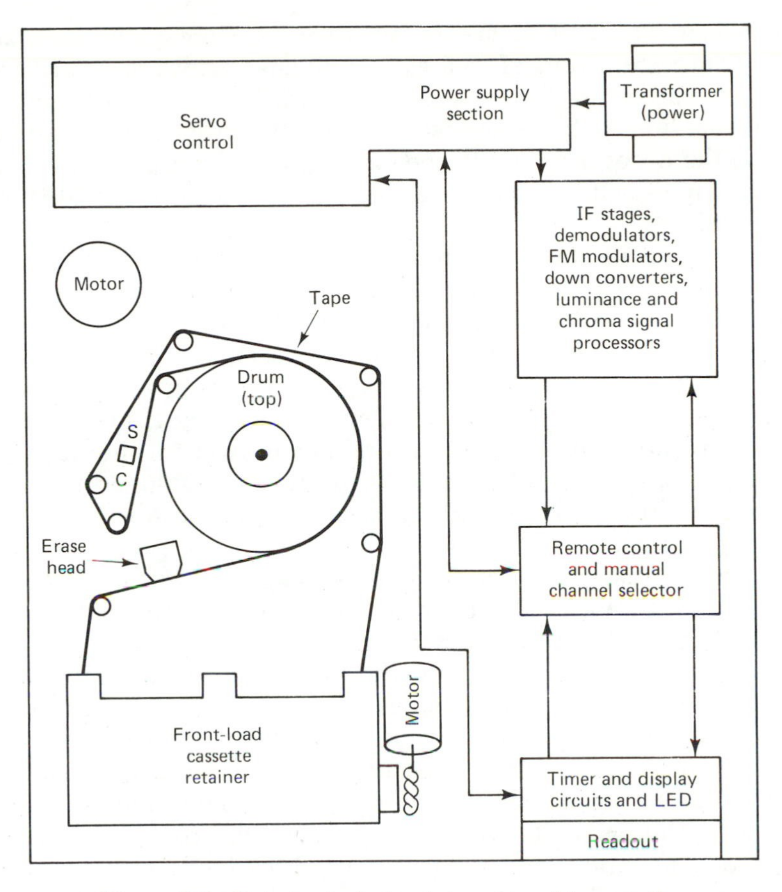

Figure 6-8 Beta typical chassis layout and tape path.

The drum containing the video recording/playback heads is at the center left, as shown. The motor for rotating the drum appears at the upper left of the drum. Another motor is shown beside the cassette retainer and is used for transporting the cassette a sufficient distance into the loading area so that the drum assembly can engage the tape. The same transporting motor performs the eject function when this is initiated by pushing the eject button. The drum containing the tape heads is surrounded by tape guideposts and fixed heads for control, erase, and audio purposes.

Because of the fixed and rotating head system as well as the slant at which the tape rides on the drum, the loading of a cassette is much more complex than is the case with audio cassettes. If the tape from the video cassette fed the tape head drum directly, it would be necessary to have one end lower than the other around the drum. With such an abrupt tape slant there would be problems in tape binding and with improper feed. Thus, VCRs are designed so that the tape is automatically loaded onto the drum as well as around the various guidepost rollers. Since the path from the cassette holder to the drum is now elongated instead of direct, the guideposts can be altered in height gradually to provide the necessary slant to the tape across the drum. In the Beta system the tape loading has a U-configuration instead of the M-type used with VHS.

The drum mount rests on a circular plastic section which is tilted at an angle. This plastic section is grooved so that the drum mount can ride in the groove. With the tape threaded in the proper manner, the appearance is shown in Fig. 6-8. When a cassette is placed into the VCR receptacle, the drum assembly is rotated and rides in the groove section of the slanted plastic case. The guideposts shown at the extreme left in Fig. 6-8 are thus rotated clockwise as they move with the entire assembly. At the extreme of the clockwise motion the guideposts are under the front of the cassette and behind the tape section that spans the front of the tape.

Next the assembly rotates counterclockwise and carries the tape section with it. During this counterclockwise rotation the tape is automatically threaded properly over the appropriate guideposts. Finally, the assembly comes to a rest in the position shown. Now, when the record or playback button is depressed, the drum spins at 1800 rpm, as discussed earlier in Sec. 6-2. Simultaneously, the tape moves from left to right and passes the erase head before it engages the video heads in the drum assembly. The tape then passes the sound and control heads as it leaves the drum area. Next the tape travels progressively along the guideposts, and the latter gradually restore the tape to the same level as it was originally. Thus, as it enters one side of the cassette and exits from the other, the tape is at the same level, and the slant only occurs when the tape wraps around the drum.

In earlier Beta video cassette recorders the tape configuration around the drum is maintained during fast-forward and rewind modes. During the latter, however, the tape tension is reduced considerably, and hence the tape pressure against the drum is much less than during playback or recording. In the later model Beta machines the tape disengages from the drum during the fast-forward and rewind modes but maintains the U-position during the review-cue modes.

7

Test Equipment Types and Usage

7-1. THE VOLT-OHMMETER

Although many basic tests and checks can be made without using test equipment, the latter is essential for advanced troubleshooting procedures. Even the most simple test equipment expedites the localization of defective components and provides a ready means for testing for the presence or absence of certain voltages. A variety of test equipment is available, though a few basic units suffice for most all necessary tests and measurements. Of considerable value is the volt-ohmmeter (VOM) unit shown in Fig. 7-1. If any test equipment is to be purchased by a newcomer, this would be the initial one to acquire. A volt-ohmmeter such as illustrated measures the ac and dc potentials encountered in VCRs and can also be used to test for open, shorted, or partially shorted components and circuits. It is not too difficult to utilize the instrument, and the more one becomes familiar with its potentials, the more invaluable it becomes in testing and troubleshooting.

Even the least expensive volt-ohmmeter can be used to good advantage in testing and troubleshooting. The moderately priced units, however, have a greater degree of sensitivity and have a wider range of features that facilitate their usage. As a voltmeter, the unit can be utilized on ac or dc within the specific ranges indicated on the meter face. When a meter is used to test voltages in a circuit; however, the meter itself may upset circuit function when it is connected across a component. Since the internal circuitry of the meter has a degree of resistance, it is

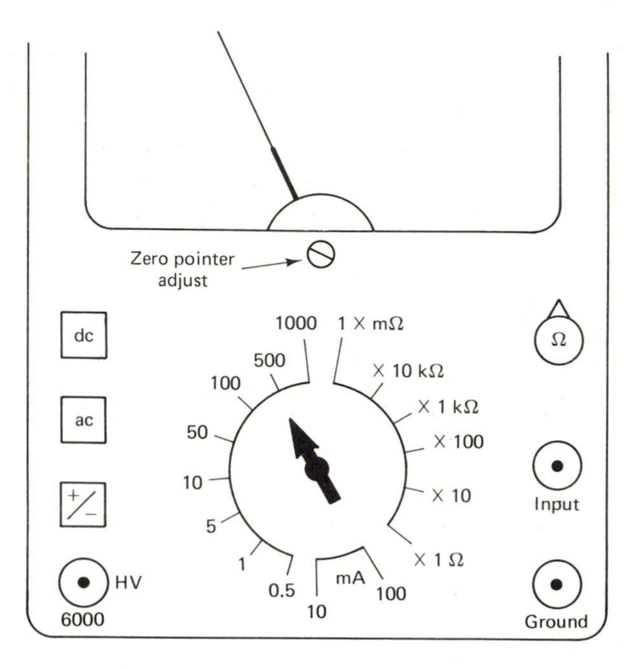

Figure 7-1 VOM selector switches.

advisable that such an internal meter resistance be as high as possible to minimize what is termed *loading effects.*

Volt-ohmmeter units are available in analog types, with scale divisions as shown in Fig. 7-2(a), and digital types, with a display face as in Fig. 7-2(b). The versatility of either instrument depends on the cost factor, with the higher-priced units having a greater sensitivity and a wider range of measuring potential. For the analog type, primary measurements consist of voltage measurements of either alternating current or direct current. Most meters also provide for direct-current readings in the milliampere regions. For a sensitive meter the range of dc current readings could encompass a span from 50 μA to 500 mA.

Another scale is for reading the value of resistance in ohms, and hence the function is that of an ohmmeter. For the latter an internal voltage (usually batteries for the portable meters) is applied through appropriate resistive circuitry. When the test probes are placed together, the maximum amount of battery voltage is applied for the particular ohmmeter range selected. Since this would cause a maximum reading, it represents zero resistance. Some ohmmeter scales read from 0 Ω at the right to a high maximum value at the left. Again, depending on the design and the sensitivity of the instrument, readings in the megohm range can be made. When the test probes are placed across a resistor, the latter permits current flow through the meter but to a

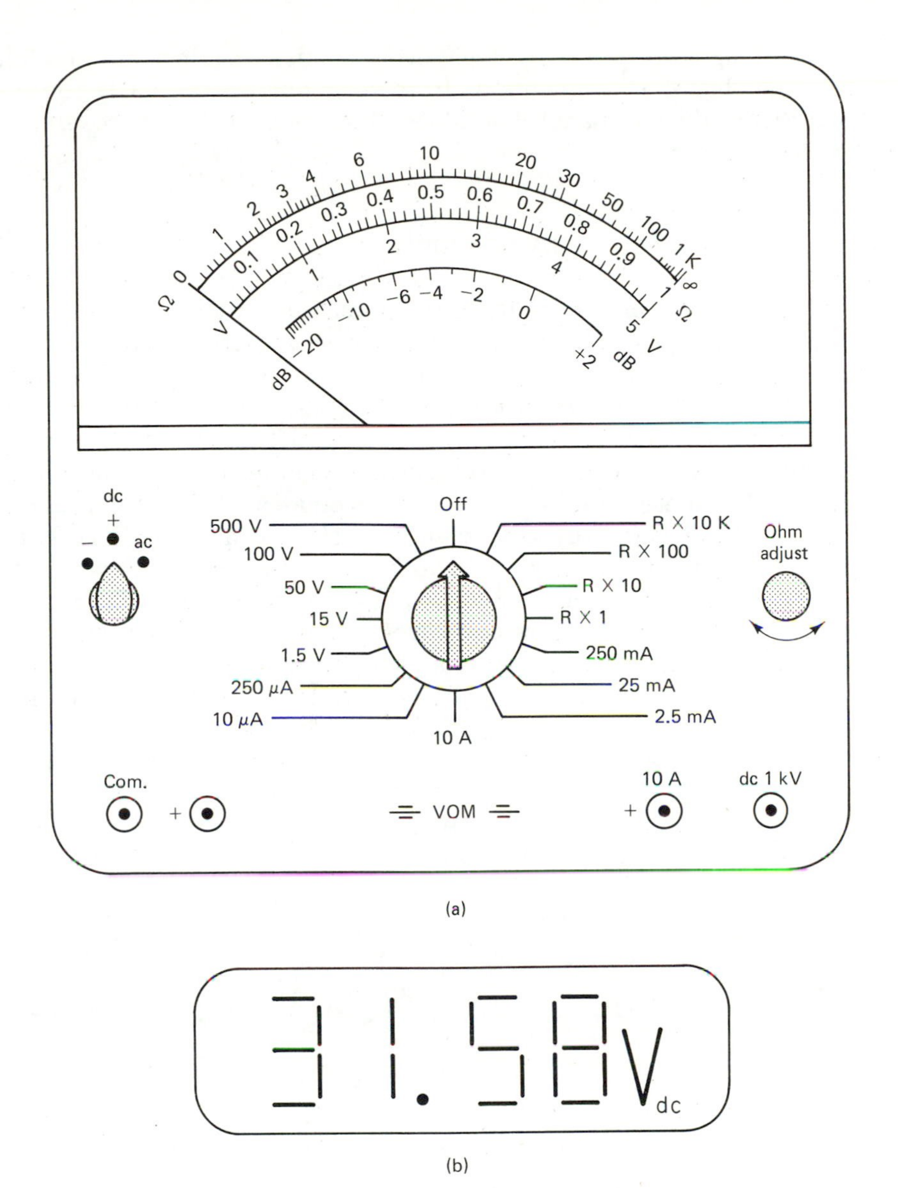

Figure 7-2 Analog and digital VOM displays.

lesser degree than when the test probes are placed together. Thus, the higher the resistance, the less current flow through the meter circuit. The needle then moves to indicate higher resistance.

The voltage, current, and ohm reading capabilities are the most widely used in normal troubleshooting practices. Some meters, however, also have a decibel (dB) scale. This scale is useful for making comparisons of signal power or voltage levels. Similar readings are possible with some digital instruments and are read out on the LCD (liquid-crystal diode) display panel. Usefulness and applications are covered in the following sections.

7-2. VOLTAGE AND CURRENT MEASUREMENTS

The ac ranges of a meter are utilized for measuring the 120-V line potentials from the power mains as well as the ac potentials obtained from the secondary windings of transformers. To avoid overload when measuring voltages obtained from the power mains, care must be taken to have the meter adjusted for the proper range. If, for instance, a test is to be made of the wall socket to ascertain whether or not line voltage is present, the meter would have to be set on an ac scale ranging above the approximate 120 V usually present, as shown in Fig. 7-3(a). Thus, a range of from 0 to 200 or 0 to 150 if present on the meter would be sufficient. In making measurements of this sort, extreme care must be exercised to keep the fingers away from the test probe points and to take all necessary precautions to avoid shock hazards.

The same ac scale range can be used to test for the presence of ac in the primary of a power transformer, as shown in Fig. 7-3(b). For measuring the secondary voltage, the test probes are placed against the output terminals, as shown in Fig. 7-3(c). If the voltage output is unknown, start with a higher-range scale and gradually reduce the range until a reading is obtained. As shown, if a 30-V output is present at the

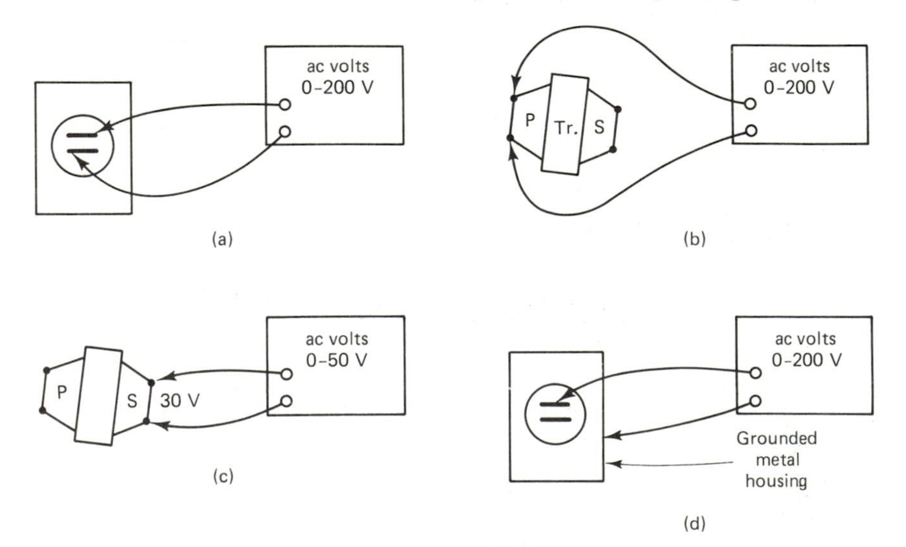

Figure 7-3 Meter scale settings (ac).

transformer, a satisfactory meter range for reading this voltage is with the ac scale set for 0 to 50. If it is necessary to establish which side of the ac line is at ground potential, place one probe on the metal junction box housing (or on a ground section such as a water pipe). The other probe can now be placed in each slot of the socket. If no reading is obtained in one slot, the latter is identified as the ground line.

Wall sockets and ac plugs should be polarized for protection against touching metal parts that have potentially dangerous line voltages present. By using a polarized plug and socket as shown in Fig. 7-4(a), the VCR is automatically grounded when the plug is inserted into the socket. Because the plug can only be placed into the socket in one position, a common ground is automatically established. An alternate method is that shown in Fig. 7-4(b) where a three-prong plug is used, with the extra plug representing the ground line.

In taking dc voltage measurements, the accuracy of the reading depends to a considerable degree on the meter's sensitivity. Meter sensitivity is related to the internal resistance of the basic meter movement or, in digital voltmeters, to the sensitivity and input impedance of the circuit fed by the test probes. In analog meters the sensitivity is known as *ohms per volt*. For such meters, successively higher voltage readings are obtained by switching higher resistances in series with the probes and the meter movement as the selector knob is turned. If a meter is rated at *one thousand ohms per volt*, it means that for a 5-V reading the meter resistance would be 5000 Ω, while at 200 V the meter resistance would be 200 kΩ (200,000 Ω).

The ohms-per-volt factor is important because during voltage tests an inferior meter would load down the circuit and hence give false readings. Assume, for instance, that the voltage across the resistor shown in Fig. 7-5 is to be measured. If 5 mA of current were flowing in the circuit through a 5000-Ω resistor, the voltage drop across the latter would be 25 V ($E = IR = 0.005 \times 5000 = 25$ V). If a meter were placed across this resistor and the internal resistance were only 1000 Ω/V at

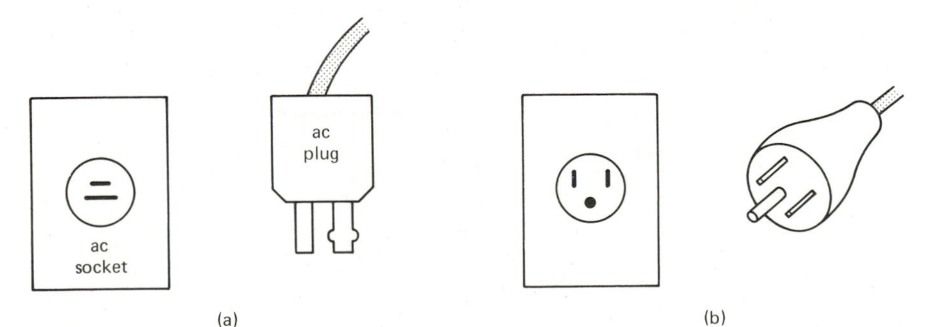

Figure 7-4 Safety plugs and sockets.

a meter scale of 0 to 25 V, the internal resistance of the meter would be 25,000 Ω. Thus, it is as though a 25,000-Ω resistor were placed across the 5000-Ω resistor R_1 of Fig. 7-5. This lowers the total value of the resistance, and hence there would no longer be a 25-V drop across R_1. Consequently, an incorrect meter reading would result. Better meters have input resistances as high as 20 kΩ/V to over 100 kΩ/V. The higher input resistances provide for greater accuracy.

The foregoing does not imply that the meters with a high degree of sensitivity (and consequently more costly) must always be employed. Often the lower-priced meters will suffice for troubleshooting purposes so long as it is understood that some inaccuracies may be present. All dc voltages throughout the VCR may vary slightly from those specified in the VCR's service notes even though voltage-regulation systems may be employed. A manufacturer may often specify a 5 or 10% tolerance in the values of certain resistors. Thus, a resistor designated as 10,000 Ω may be satisfactory if it is actually only 9000 Ω or if it is 11,000 Ω. Obviously, then, a designated value of 10 V across this resistor would also be subject to the same tolerance variations. The degree of variation, however, provides a clue to whether or not a defect exists. If the voltage drop across a resistor is supposed to be 10 V, there is obviously something wrong if only 5 V appear across it. If, however, the reading were 9 V or 11 V, the circuit may be normal. We would be suspicious if a voltage differential was present of several volts above or below the normal tolerance. For larger voltages, however, the actual differential would be much greater in voltage though not necessarily in percentage. Thus, if the reading is indicated as 50, a 5% tolerance would mean a normal tolerance range between 45 and 55. If, however, the voltage were 35 or 40 V, it would indicate some defects. Similarly, a reading of 60 V or more would also indicate circuit problems.

Current measurements are only made infrequently because it is not often necessary to get an exact reading of current flow in a given circuit. Where voltage measurements are made by simply placing the probes

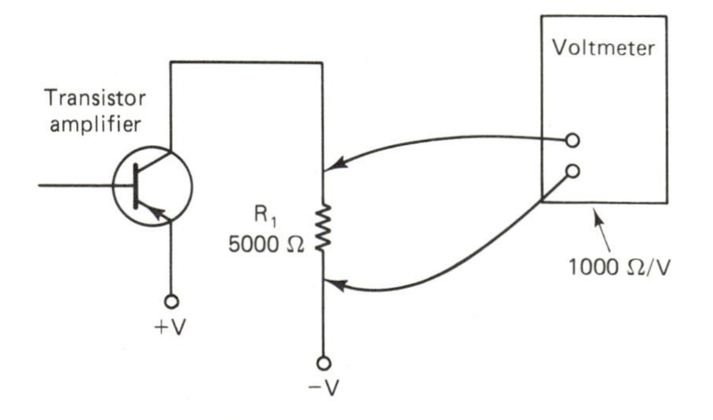

Figure 7-5 Loading effects of meter.

across the voltage source, current measurements require the opening of the circuit because the current flow must go through the meter for measuring purposes. Thus, if the dc current flow in a power supply were to be read as shown in Fig. 7-6, one section of the line is opened as shown and the device turned on for normal operation. The current is then read directly. Again, precautions should be taken by starting out with the highest meter range and decreasing the range until an appropriate reading is obtained. If the meter is on a 0–10 mA range, for instance, and 200 mA flows, the meter may be damaged. Current readings are usually avoided since the current can be calculated in many instances, if needed. Assume, for instance, that the current meter in Fig. 7-5 read 5 mA. This would indicate that 0.005 A of current is flowing. When this current flows through a 50-Ω resistor, the voltage (E times I) equals 0.25 V. Had this voltage been read across resistor R_1 in Fig. 7-6, we would have found current by dividing voltage by resistance, or 0.25/50 = 0.005 A.

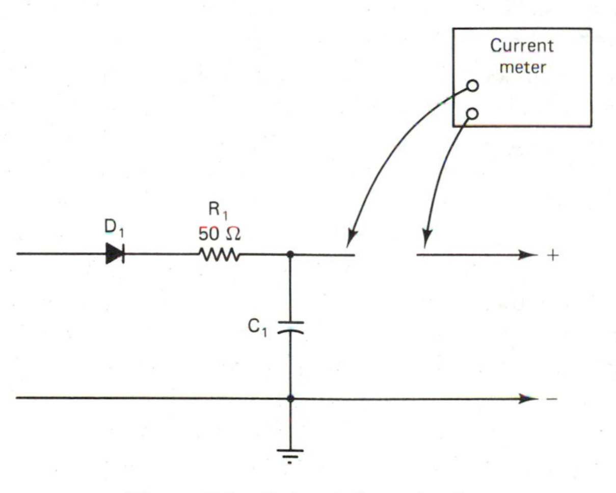

Figure 7-6 Current flow check.

7-3. OHMMETER USAGE

The ohmmeter range of a VOM, as the name implies, is useful for reading the ohmic value of resistors or the resistances of other components and circuits. Since the meter indicator movement is actuated by an internal voltage, the degree by which current flows through the circuit establishes the degree of resistance read on the dial. Often a balancing adjustment knob is present to set the meter indicator needle at true zero. This is necessary to compensate for battery aging and consequent change of calibration accuracy.

As mentioned earlier, when the probes are placed together, the reading would be 0 resistance. If, in the ohmmeter scale, the probes are separated, it indicates an infinite resistance, and the needle would swing to the opposite place on the dial. This factor makes the ohmmeter extremely useful in continuity checking of circuits, as discussed more fully later in this section.

In reading the ohmic value of a resistor, the ohmmeter probes are placed directly across the resistor, and the appropriate scale is selected so the needle deflection is not an extreme high or low. For an accurate reading of a resistor wired within a circuit, one side should be disconnected, as shown in Fig. 7-7. This is necessary because the complete circuit of the resistor may consist of other resistors in shunt. The aggregate of all these resistances would upset the meter reading. In addition to other shunting resistors, some transistors, coils, or other components may also be in shunt and hence give false readings. Some in-circuit tests can be made for a general appraisal regarding continuity, as discussed later. For an accurate reading in ohms, however, it is preferable to disconnect one side of the resistor as shown.

An important application of the ohmmeter is its use as a continuity-checking device. As such it is useful for identifying open or short circuits and open or shorted component parts. A typical test procedure consists of checking for continuity through a coaxial cable system. (See Chapter 2.) If a VCR is properly connected to a television receiver but there is no signal transfer, the first procedure is to test the coaxial cable interconnection. As shown in Fig. 7-8(a), the coaxial cable is disconnected from both the VCR and the television receiver. Next the ohmmeter probes are placed across the inner and outer conductors at one end. If any continuity is indicated, it shows that the cable is shorted internally and must be replaced. If no reading is obtained, the test probes can be placed at the input and output sections of the inner conductor, as

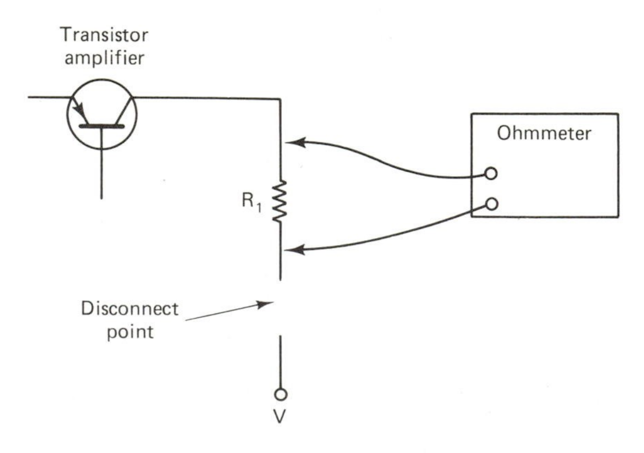

Figure 7-7 Out-of-circuit test.

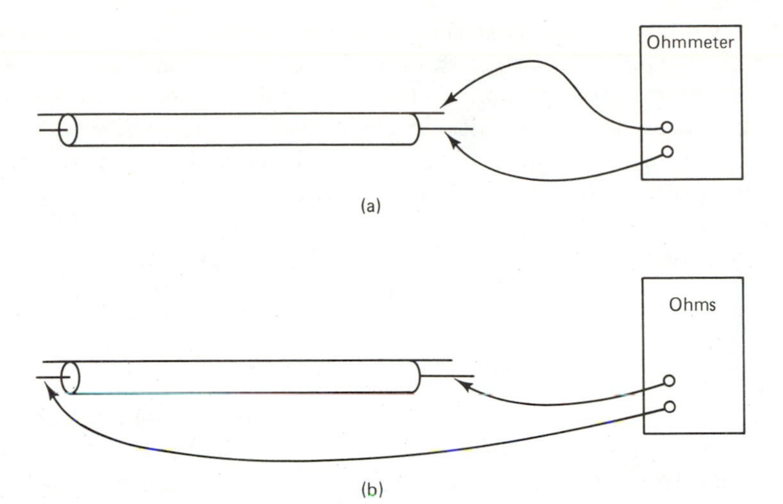

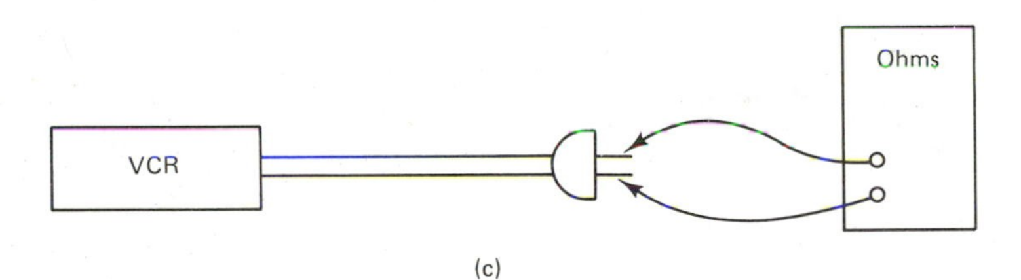

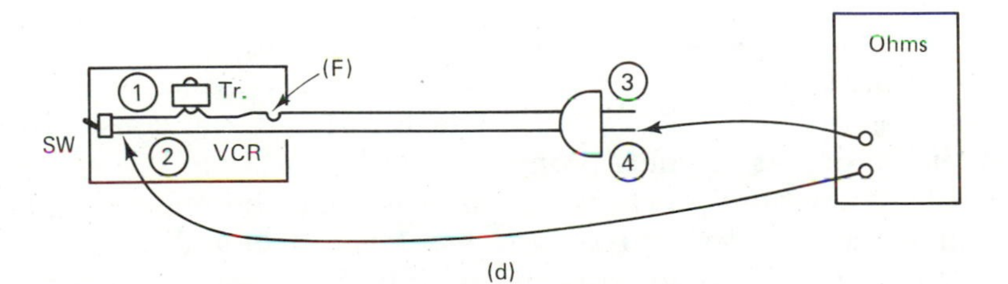

Figure 7-8 Cable and line cord checks.

shown in Fig. 7-8(b). Here continuity should be obtained since lack of continuity would indicate an open circuit for the inner conductor.

The primary purpose for continuity checking is to ascertain whether or not continuity exists between specific points in a section or a component. When using the ohmmeter as a continuity-checking device, the specific ohmic indication is of no importance. In measuring the coaxial cable as mentioned above, a sensitive meter may indicate a fraction of an ohm of resistance for a short length of cable, but so long as continuity is established between each end of the inner conductor, we have found that the cable can conduct the signal. Similarly, if continuity exists between the inner and outer conductors, we know that an integral short exists, and the cable will not transfer a signal.

In testing for continuity in electrolyte capacitors, we should obtain good continuity in one direction and very little continuity when the test probes are reversed. A continuity check for the electrolytic capacitors is useful for identifying a shorted capacitor wherein a very low resistance is obtained for both positions of the test probes. If the continuity consists of very high resistance for both directions of the test probes, it indicates a unit that is almost an open circuit within the electrolytic and hence defective. An infinitely high resistance (no continuity) indicates a definite open capacitor.

A similar continuity test applies to diode rectifiers. We should obtain a good continuity (low resistance) in one direction and a poor continuity (high resistance) when the test probes are reversed. Again, an extremely high resistive reading or a very low reading for both positions of the test probes would indicate a faulty unit.

Another useful continuity check involves the ac line cord to the VCR. If the VCR does not turn on when the on-off switch is activated, it may indicate a faulty switch or line cord. Before testing these, however, make sure that there is ac present in the wall socket by plugging in a lamp. If the lamp lights, the next procedure is a continuity check of the line cord, as shown in Fig. 7-8(c). Here no continuity should be shown when the switch is in the *off* position, but some continuity should be present when the switch is flipped to the *on* position. No continuity will show either a defective plug, cord, or switch. If the underside of the chassis is exposed, an additional continuity check can be made, as shown in Fig. 7-8(d). The initial test can consist of one ohmmeter probe at the plug prong (4) and the other at the switch terminal (2). A continuity here would indicate that the line and plug are both in working order. A test between 1 and 3 should also give continuity, though 100 Ω or so would be introduced by the series transformer primary. If no reading is obtained, the fuse (F) could be open and should be checked. If the fuse is all right, the cord or the transformer primary may be open. If a reading is obtained between points 1 and 3, a final check can be made between points 1 and 2. When the switch is thrown to the off position, there should be no continuity between these two points, and if the switch is not defective, a continuity reading should result when the switch is turned on.

A familiarization with the basic characteristics of a device is useful in making continuity checks. If problems appear to rest with the VCR power transformer, for instance, the readings obtained across the secondary winding would appear to indicate shorted conditions, but actually the transformer secondary may be all right. The apparent short-circuit conditions are indicated because the secondary winding contains fewer turns than the primary. Fewer coil turns reduce resistance, and since the secondary also uses heavier-gauge wire, the resistance is reduced additionally. Thus, placing test probes across points 4 and 5 of Fig. 7-9 may only indicate a few ohms of resistance and yet be a normal condition. For a reading between 3 and 4 the resistance would be somewhat higher but may still be only about 5 Ω and thus simulate a short circuit when actually it is a normal reading. So far as continuity goes, it can be assumed the transformer secondary winding is all right unless the meter reads 0 Ω on the lowest-ohm scale during the continuity testing. It would also be advisable to test the voltage across each secondary section with the power on. Make sure *no continuity tests are undertaken* when power is present in the transformer.

In testing the primary coil section across points 1 and 2, the continuity should not show zero resistance but rather 50 to 100 Ω or so depending on design. If the continuity here is so high the ohmmeter reads zero, it is probable that a short circuit exists in the noise filter capacitor C_1. It is rare for the primary to be shorted completely unless the transformer had burned out and all the insulation of the primary coil had been destroyed.

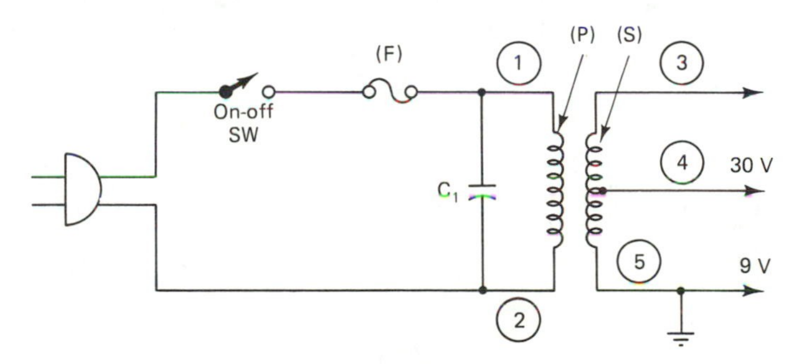

Figure 7-9 Transformer tests.

7-4. THE OSCILLOSCOPE

The oscilloscope is a test instrument that enables us to obtain picture images of the actual signal within a circuit. The common term of the instrument is usually the shortened word *scope*. Although the scope is not an instrument necessary for normal troubleshooting procedures, an understanding of its applications enables us to expedite the location of faulty components. The scope indicates the presence or absence of the required signal, whether such a signal has the proper amplitude and whether or not the signal has been distorted by some circuit problem. The scope uses a cathode-ray tube that functions in a similar fashion to that of a television tube. Both have a phosphor-coated face plate which fluoresces when struck by an internal electron beam. In this fashion images are traced out on the face plate of the tube. Although conceivably television pictures could be produced on the scope screen, its design is such that it only displays ac signals (both audio- and radio-frequency tapes) and amplitudes of dc signals.

The scope produces an electron beam by releasing energy from its cathode structure and accelerating it to the phosphor face plate. To show a single sine-wave display, as illustrated in Fig. 7-10, requires the scope to move the beam from left to right once each time the sine-wave signal is obtained by the input test probes. For a 60-Hz signal the internal beam is swept across the face of the tube once for each cycle. If the beam travels across the tube face at one-half the speed, two sine waves would appear. For a horizontal sweep (scan) of twice that of the sine-wave frequency, only one alternation would appear on the screen. To

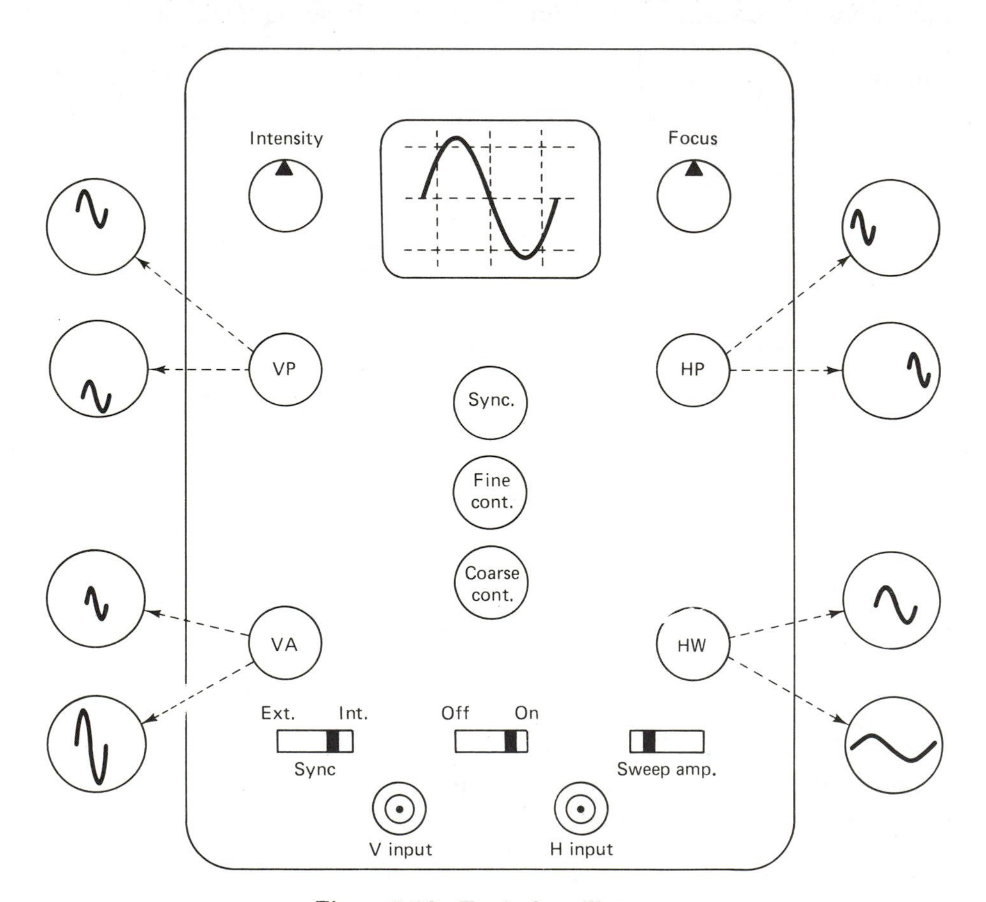

Figure 7-10 Typical oscilloscope.

generate such a scan, an internal sweep signal oscillator is employed. The latter's frequency is controlled by a front-panel knob so we can display the type of signal needed. Controls are also present for amplifying the signal vertically or horizontally to show different aspects of the waveform as needed. In general testing procedures the signal is picked up by the vertical input terminal and ground as shown. In some complex laboratory test procedures the internal sweep oscillator's signal can be replaced by an external signal input for special displays. For VCR servicing such procedures are not required. If more information is needed, reference should be made to the operator's manual that accompanies the oscilloscope. A full discussion of the various aspects of oscilloscope applications and functions would require a separate text.

Figure 7-10 depicts the effect each control knob has on a signal. Here a single sine-wave signal is being displayed on the screen. By manipulating the controls, the signal can be shifted up or down as well as to the left and right. Similarly, it can be elongated vertically or horizontally as required.

A representative low-frequency waveform is shown in Fig. 7-11. Audio signals would consist of a complex waveform rather than a single line sequence. Square waves, as shown in Fig. 7-12, can be displayed for testing audio response. Signal generators (of the type discussed later) can inject a pure square wave in an audio system, and if the scope shows a distortion of the signal at a later stage, the trouble source is then isolated. The presence of the video signal in the antenna input sec-

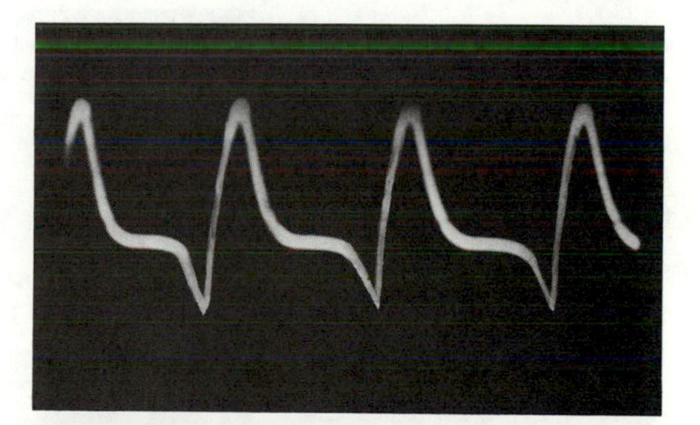

Figure 7-11 Low-frequency waveform.

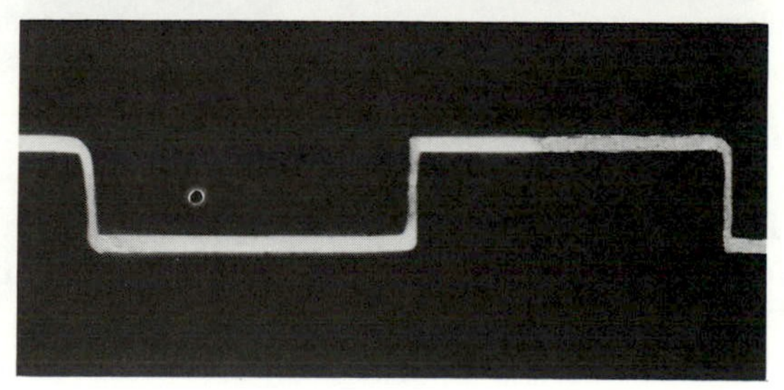

Figure 7-12 Square waves.

tions of the VCR or in the internal circuit can be checked with the scope. A typical pattern of a television signal is shown in Fig. 7-13. Here the synchronizing pulses are at the peak of the waveform, while the picture information appears between the peak points.

The more costly oscilloscopes are capable of displaying two waveforms at once. Thus, for VCR systems incorporating stereo, each channel can be checked and compared to the other to ascertain whether or not power output and purity of waveforms coincide. A typical dual-trace scope display is shown in Fig. 7-14.

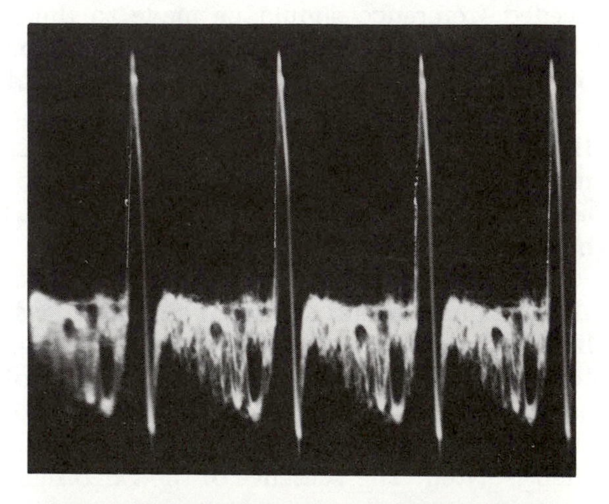

Figure 7-13 TV signal pattern (RF).

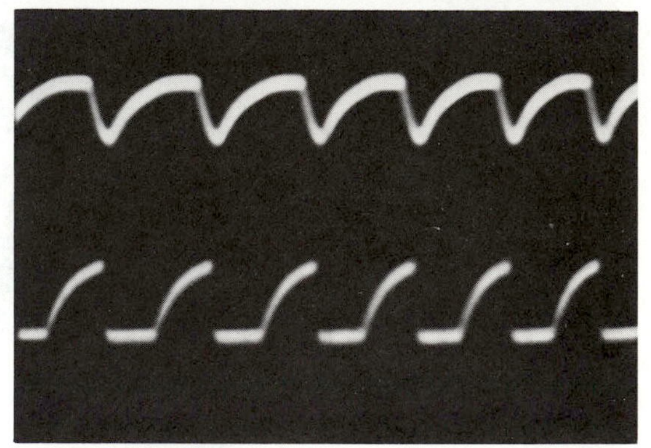

Figure 7-14 Dual trace display.

7-5. SCOPE APPLICATIONS

The versatility of the oscilloscope lies in its ability to produce a visual image of virtually any signal waveform found in electronics. As mentioned in Sec. 7-4, the displayed image can be amplified to expand the signal either vertically or horizontally. This enables us to view a reasonably sized signal on the screen. Thus, the signal viewed may have an amplitude that differs from the original, but the display is still the basic signal waveform. While the height and width controls may enlarge the viewed image, they do not add signal distortion. In electronics, the term *distortion* implies an alteration of a waveform which could include an amplitude increase of the positive alternation of a sine wave followed by a decrease in the negative alternation, as shown in Fig. 7-15(a). If the pure sine wave were symmetrically altered (where both alternations increase or decrease simultaneously) such as in Fig. 7-15(b) or elongated as shown in Fig. 7-15(c), no distortion would occur.

Other forms of distortion causes one alternation to have a longer duration than the other, as shown in Fig. 7-15(d), but amplitudes remain the same. Similarly, a distortion such as shown in Fig. 7-15(e) represents a third harmonic component because it consists of a dual train of sine waves that could be represented as shown in Fig. 7-15(f). Here the peak amplitudes of the sine waves have been altered as shown in Fig. 7-15(e), and thus additional frequencies were generated by the distortion. Other forms of distortion are shown in Fig. 7-15(g) and (h). In Fig. 7-15(g) a distortion occurs in the incline of the ac waveform, while in Fig. 7-15(h) a distortion occurred in both incline and decline. (See Sec. 1-1 and Fig. 1-1.)

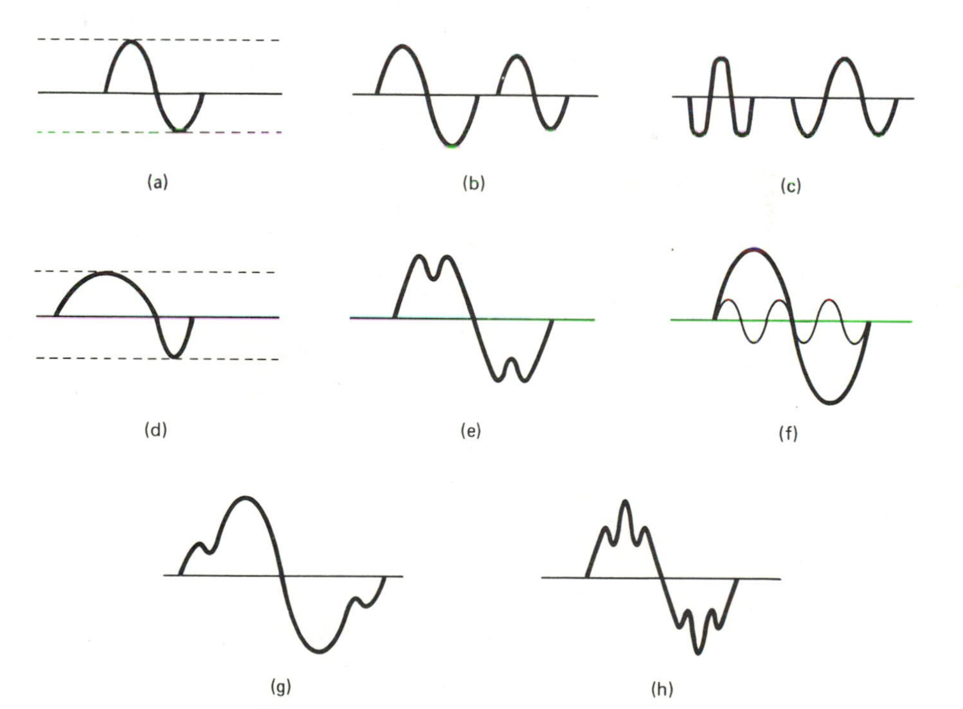

Figure 7-15 Types of signal distortion.

The foregoing discussion on signal distortion emphasizes the value of the oscilloscope because the latter permits us to test for such distortion and correct it as required. As an illustration we can inject a pure sine-wave signal into an audio amplifier system, as shown in Fig. 7-16(a). By placing an oscilloscope at the output of the amplifier, we should get an amplified version of the original sine wave. If, however, the sine wave appears in any of the distorted forms shown in Fig. 7-15, it would indicate a defective component or circuit in the audio stages. For this test the horizontal sweep frequency of the oscilloscope is set at a low frequency (in the audio range) for proper sweep function. A similar

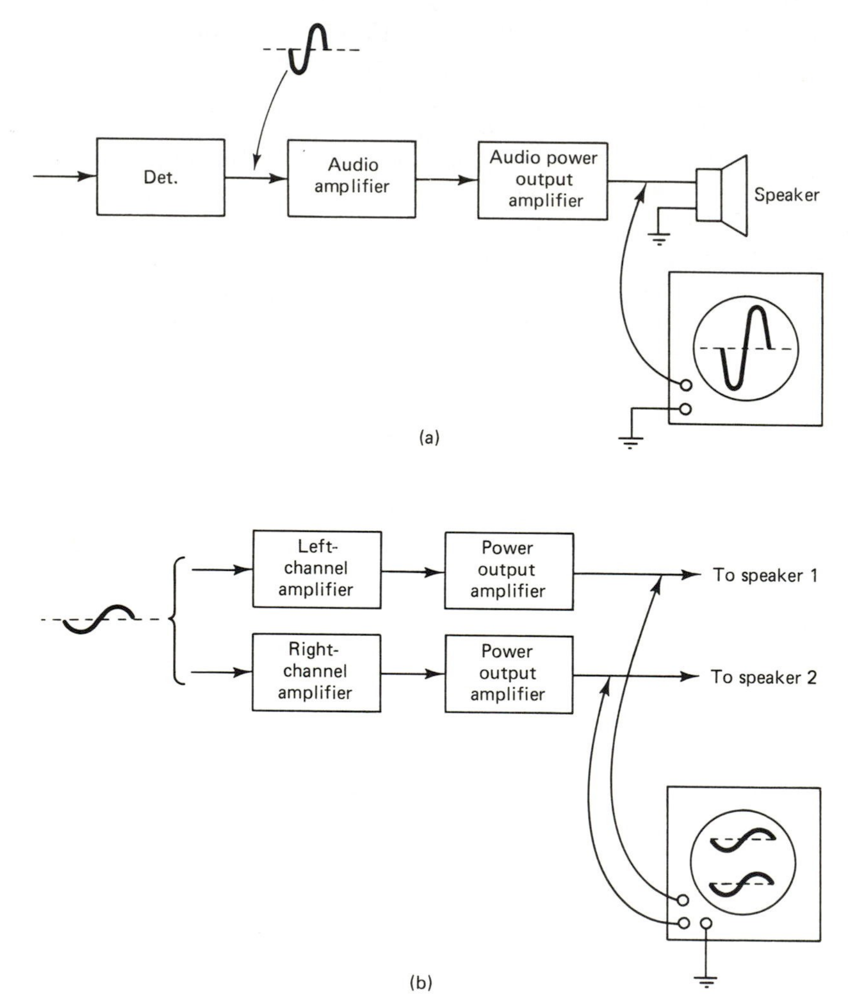

Figure 7-16 Signal tests.

test could be performed on RF or IF stages with appropriate changes in the horizontal sweep frequency, as discussed more fully in Chapter 10.

For a stereo system, a similar test procedure would be utilized for both the left and the right channels. With a dual-trace scope, a signal is simultaneously injected into both right and left amplifiers and the dual output viewed on the screen, as shown in Fig. 7-16(b). Such a procedure is useful for testing whether or not one channel produces more distortion than the other. If the balance control of the stereo system is set at center and the gain controls in the scope are set at the same level, identical waveforms should appear on the screen. A square wave can also be injected and the output checked for an exact reproduction. Any deviation from the square waveform indicates distortion.

In many electronic circuits the ac type of signal representing audio, RF, or video is referenced in peak-to-peak voltage values rather than the conventional rms values. (See Fig. 1-1.) The peak-to-peak value represents the voltage between the peak of the positive alternation and the peak of the negative alternation. Thus, if the rms value read on a conventional voltmeter is 42.4 V, the peak value would be approximately 60 V for a single alternation. Thus, the peak-to-peak voltage (P-P) would be 120.

The scope is useful for measuring such peak-to-peak values and is easily calibrated, as shown in Fig. 7-17. In Fig. 7-17(a) is a conventional sine wave representing the peak-to-peak value. A transparent plastic sheet layed out in graph form is placed over the oscilloscope face plate as shown in Fig. 7-17(b) and provides calibration reference points. The horizontal sweep of the scope is shut off, and the sine-wave image then produces a single vertical line as shown. If the peak-to-peak voltage were in the 100-V region, each square of the face plate graph can be allocated a value of 10 V. The incoming signal is then measured with a conventional voltmeter and the amplitude of the display altered by the vertical gain control so that 100 V occupies exactly 10 spaces, as shown. Consequently, any peak-to-peak voltage of 120 V or less will be accu-

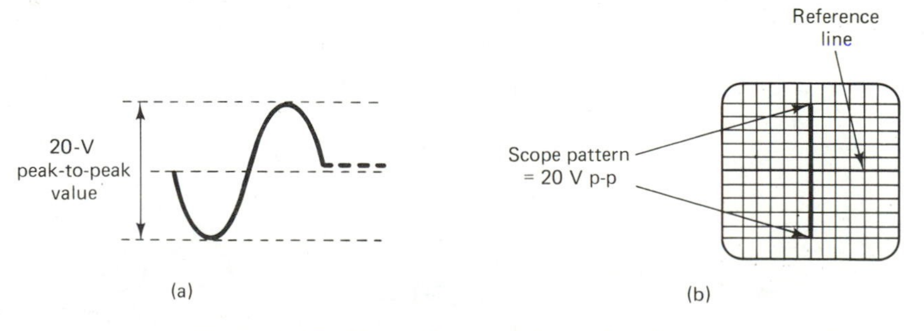

Figure 7-17 Measuring peak-to-peak voltages.

rately displayed on the screen. For higher voltages the screen would have to be recalibrated accordingly.

The oscilloscope is also useful for signal tracing purposes as discussed in Chapter 10 and also for testing for the absence or presence of signals to aid in defective-component localization techniques, as discussed later. As shown in Sec. 7-8, the oscilloscope system is also used in the color signal tester known as a vectorscope.

7-6. SIGNAL GENERATORS

Signal generators are useful devices in troubleshooting procedures because they produce signals that can be injected into circuits to check for amplification and signal progression through electronic circuits. A basic signal generator is shown in Fig. 7-18. This device forms a single signal, and either a sine wave or a square wave can be selected by the push-button control at the upper left. The signal can also be modulated by a 400-Hz signal for testing RF and detector stages. An output level control permits adjustment of the signal amplitude obtained at the output. A range switch is used to select signal frequencies beginning at the lower audio levels and extending into the upper radio-frequency range. The exact frequency of the output signal is selected by the central dial.

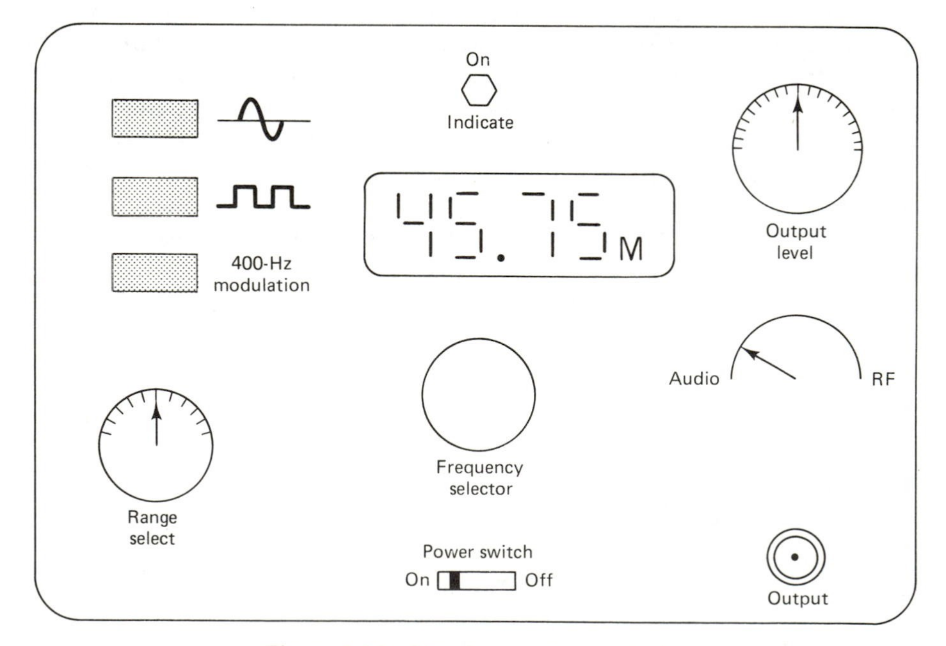

Figure 7-18 Signal generator controls.

Some instruments of this type will display the exact frequency selected in a digital readout fashion. The signal generator is also useful in signal tracing purposes, as discussed in Chapter 10. Modern devices of this type have good frequency stability and acceptably accurate frequency control.

Another type of signal generator is one in which the output signal is varied in frequency over a selected range. Such a signal generator is termed a *sweep generator* because of the signal sweep above and below a selected frequency. Such a sweep generator has a number of controls that regulate the center frequency, the rate of sweep, and other manual adjustable functions necessary for correct operation. A typical front-panel layout may have the configuration shown in Fig. 7-19(a), though all these controls may have different positions depending on the manufacturer.

When a sweep generator signal is injected into a circuit, it can form visual display of the frequency response of an IF or RF stage on the oscilloscope screen. If, for instance, the sweep generator injects a signal into the IF amplifier circuits, an oscilloscope placed at the output would produce a pattern as shown in Fig. 7-19(b). This would show the

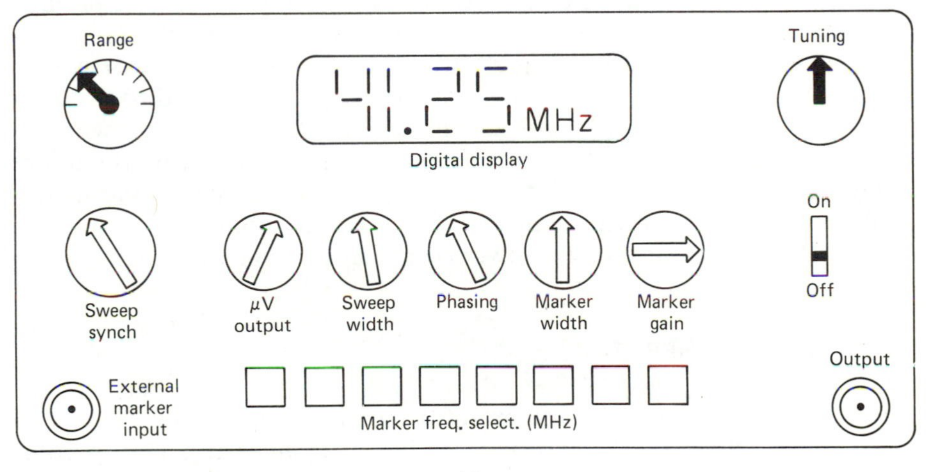

(a)

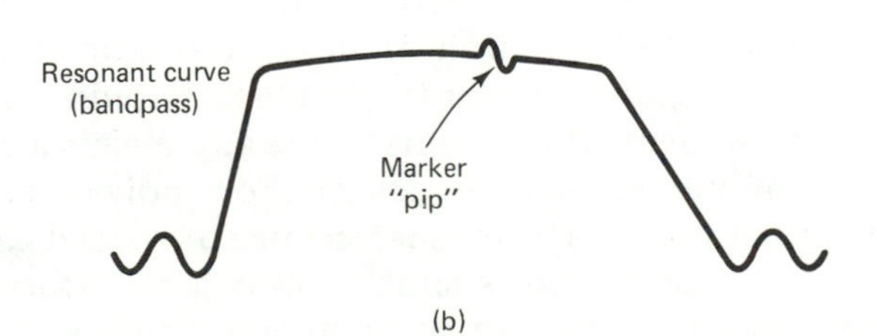

(b)

Figure 7-19 Sweep generator and marker pattern.

exact shape of the bandpass plus the specific dips that occur because of the traps utilized, as shown earlier in Fig. 1-10(b). As the generator sweeps the signal frequency below and above the resonant frequency of the IF stages, the signal displays the relative gain along the bandpass curve but does not indicate the exact frequency at any given point. Consequently, an additional oscillator (producing a single frequency) is utilized within the sweep generator to produce an accurate frequency marker at any point selected on the bandpass curve. The single signal (mixed with the sweep frequency signal) produces a break in the pattern along the response curve. This break is termed a *marker pip*. By having a series of marker frequencies under accurate control, different ones can be selected to indicate the exact frequency at points along the response curve. By pushing various marker-frequency buttons as shown in Fig. 7-19(a), the marker pip can be moved along the response curve to determine the exact frequency span of the IF bandpass displayed.

As shown in Fig. 7-19(a), various controls permit exact adjustment of the display pattern. The sweep range is determined by one control, the output in microvolts by another control, and the sweep width by still another control, as shown. A sweep synchronization control is available which, in conjunction with the phase control, permits a stable lock-in of the visual pattern. The marker pip width and amplitude can also be controlled as shown. If an external generator with a stable and accurate frequency is available, it can be utilized to produce the marker by injecting its output into the lower left input shown. The tuning of the center of the sweep frequency is performed by the upper right dial, and the visual display in megahertz is shown in the meter-dial face plate above the controls. The coaxial cable output for this particular instrument is at the lower right-hand corner, as shown.

7-7. PATTERN GENERATORS

In the adjustment of VCR and monitor receivers it is often necessary to correct for faulty color renditions or incorrect linearity. Faulty color may be the result of a misadjustment of controls rather than one of defective circuitry. Hence, color deterioration can be caused by mistuned sections in the VCR as well as in the monitor or television receiver. In the latter instance, linearity problems should not be present often since an adjustment of the controls usually means a good holding pattern because of modern circuit design. For individuals who, however, wish to obtain peak performance, the use of special generators for the adjustment of color monitors and receivers is necessary. One of the most common types is a generator combining bar patterns with dot and color patterns for adjustment of both linearity and color balance. Thus, such a device is called a *crosshatch and bar* generator and is illustrated in Fig. 7-20.

This generator forms vertical and horizontal bars on the television screen for adjustment of both vertical and horizontal linearity. For a display of vertical bars, for instance, there should be equal spacing across the picture tube. For any crowding of the bars on one side as opposed to the other, horizontal nonlinearity would be indicated. Similarly, when the horizontal bar pattern is injected into the antenna system of a receiver, the series of horizontal bars appearing on the screen should be evenly spaced from top to bottom. Any crowding at the upper or lower end again indicates nonlinearity and should be corrected by adjustment of the vertical linearity control. A sample bar pattern for checking vertical linearity is shown in Fig. 7-21.

A crosshatch pattern permits a check of both the vertical and

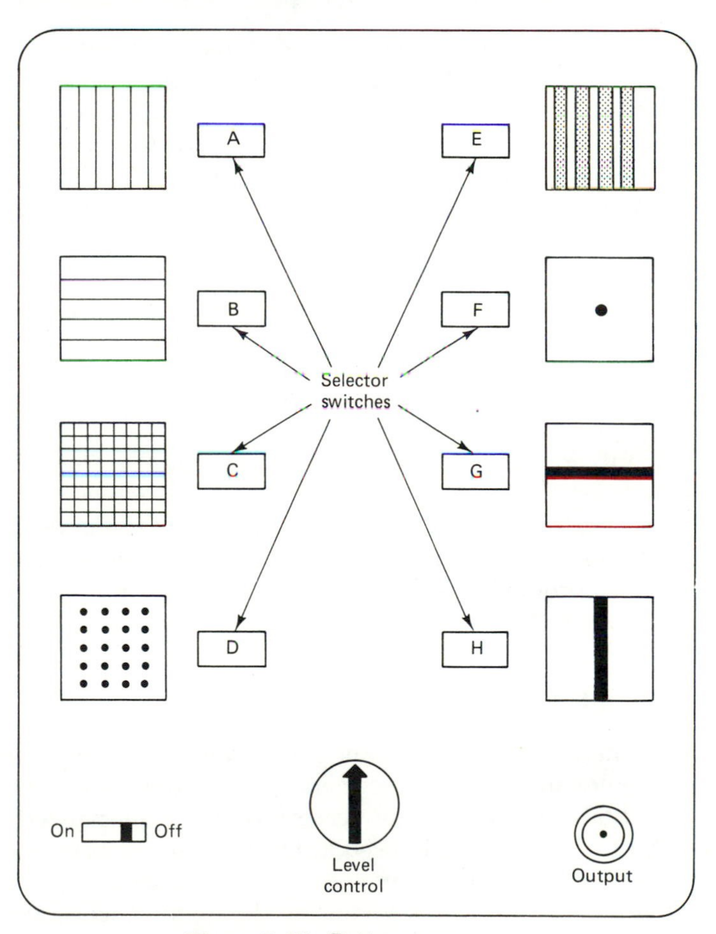

Figure 7-20 Pattern generator.

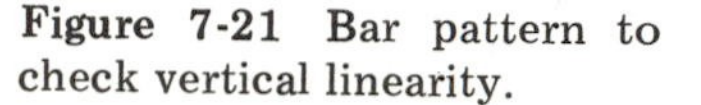

Figure 7-21 Bar pattern to check vertical linearity.

horizontal linearity at the same time. A single line (either vertical or horizontal) is available in most meters, and the line can be positioned either at the top or bottom as well as left or right for testing linearity distortion. The latter causes curvature of a straight line. The other patterns available are the dot pattern and the color-bar pattern. The dot pattern is useful for adjusting proper color rendition. A series of uniform white dots should appear if the conversion controls have been properly adjusted. As mentioned earlier, it is the proper merging of the red, blue, and green electron beams on each phosphor dot on the tube face that produces a pure white pattern. Hence, the conversion controls are adjusted to make corrections at the right or left and top or bottom until a pure white dot pattern is obtained.

The vertical color bars are produced on the screen when the appropriate push button (E) is depressed, as shown in Fig. 7-22. The color bars are used for testing the phase relationships between the several color signals, and they also aid in the adjustment of the individual color signal amplitudes. On occasion a special instrument is utilized for the sole purpose of producing such color bars and is sometimes termed a *color-bar generator* or a *rainbow generator.*

7-8. THE VECTORSCOPE

Another test instrument useful in checking and adjusting for accurate renditions of color in television receivers or monitors is the *vectorscope.* As with the dual-trace scope, the vectorscope is a specialized piece of equipment that is not as widely used as the basic units discussed earlier. Most of the color adjustments, including tint and balance, can be performed satisfactorily without resorting to the highly specialized devices such as the vectorscope and the dual-trace triggered sweep oscilloscope.

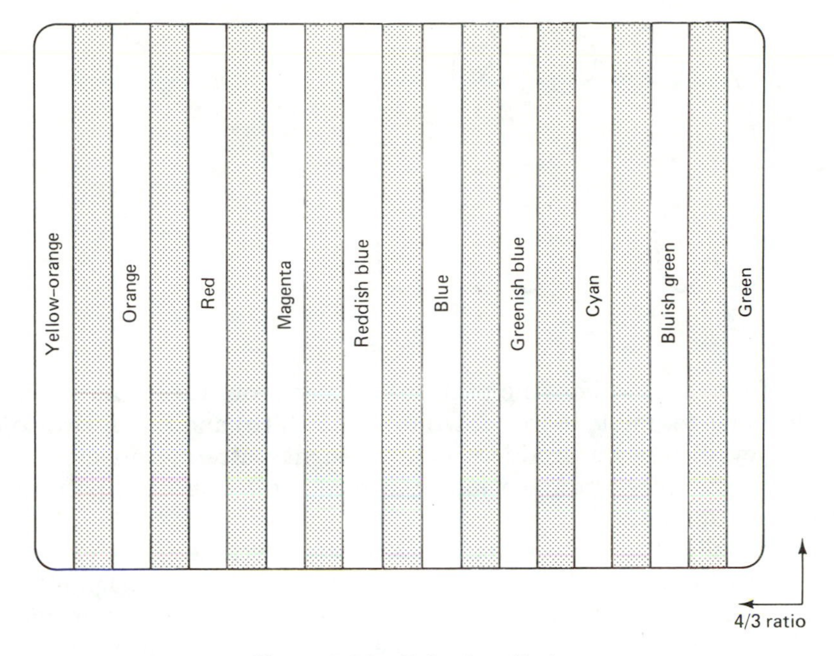

Figure 7-22 Color-bar display.

Since a discussion of the vectorscope rounds out the general equipment encountered in adjustments of color receivers and monitors, a brief outline of its function and application is included herein. As with other test equipment described in this chapter, the operating instructions which accompany a vectorscope when purchased furnish complete application details for the particular model in question.

An understanding of the functional aspects of the vectorscope is facilitated by a knowledge of phase patterns of dual signals in a scope. When we turn off the internal horizontal sweep generator of a scope and apply pure sine-wave signals to both the vertical and horizontal input jacks, the vertical sine-wave input will be moved across the screen by the horizontal sine-wave input, which influences the lateral beam movement. Consequently, a series of patterns forms on the screen, and their waveshape depends on the signal frequency and amplitudes of the input signals. Basic waveshapes of dual sine-wave signal inputs are shown in Fig. 7-23, and these are termed *Lissajous* patterns after the famous French scientist who experimented in frequency differences of physical movement. When two signals have the same frequency but one is 90° out of phase with the other, a circle appears on the scope screen. If there is a 135° difference between the signals, an oval appears, while for a 45° phase difference, the oval is tilted in the other direction, as shown. When the two-input sine-wave signals have frequencies that

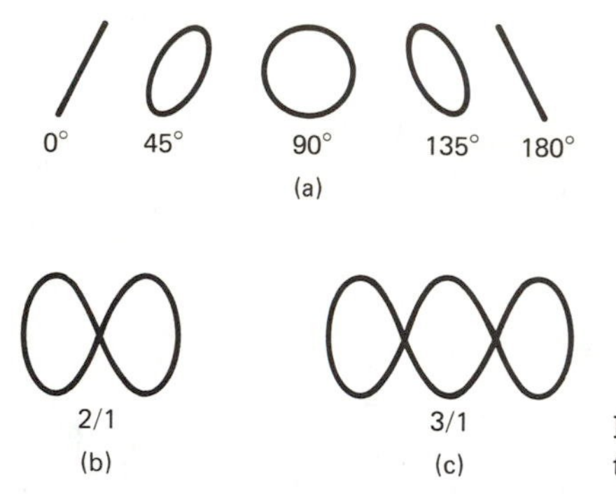

Figure 7-23 Lissajous scope patterns.

differ from each other, the patterns change from those shown in Fig. 7-23(a) and resemble those shown in Fig. 7-23(b) and (c). A two-to-one frequency difference produces a figure-eight pattern, and greater frequency differences produce multiple oval interconnected patterns, as shown in Fig. 7-23(c).

Basically, the vectorscope is essentially a well-designed scope using a special face plate with markings related to the vector diagram shown in Fig. 7-24. In this diagram the various hues in the television receiver are illustrated. The hues are produced by changing the phase of the color subcarrier with reference to the burst signal. Hence, if the phase

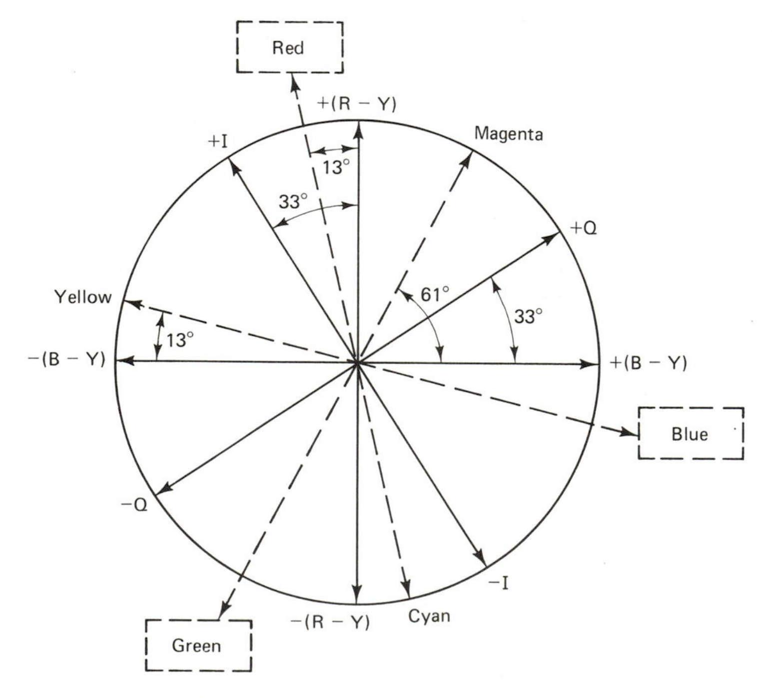

Figure 7-24 Color TV signal phases.

of the subcarrier signal were changed continuously through 360° for a time duration of one horizontal line on the TV screen and repeated for each following horizontal line, a rainbow color pattern would appear. (See also Fig. 7-22.) Some vectorscope face plates have numbered segments corresponding to the 10 standard color bars. Some face plates have been used with segments identified by their respective vector angles. A separate rainbow generator can be used, or the vectorscope can have one built in. Typical connections for testing the color demodulators are shown in Fig. 7-25.

With the test setup, the $R - Y$ demodulator signal output is applied to the vertical input of the vectorscope and the $B - Y$ to the horizontal (sweep) input. With these connections a vectorscope petal formation appears. In Fig. 7-26(a) the phase relationship of the signals

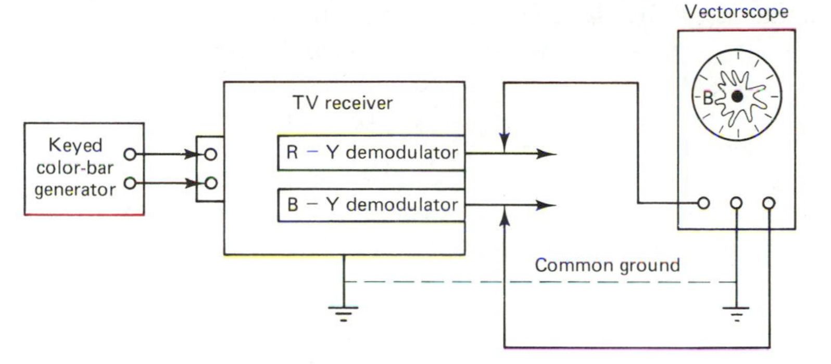

Figure 7-25 Vectorscope connections for testing.

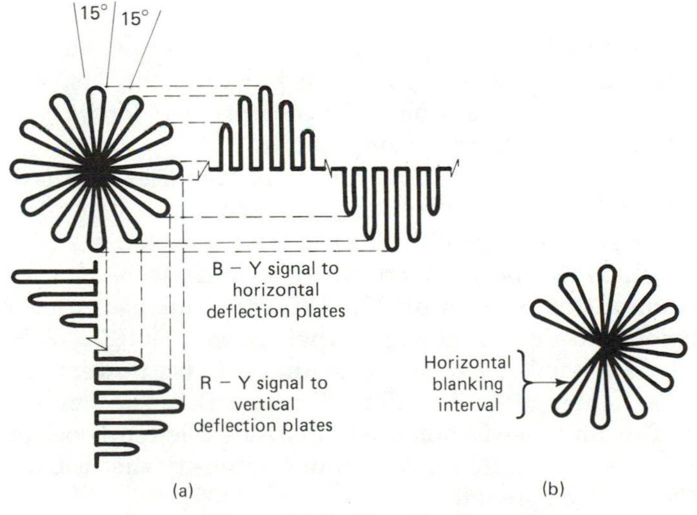

Figure 7-26 Vectorgram.

and the petal formation are shown. In Fig. 7-26(b) is shown the pattern that appears on the vectorscope face plate. This is an ideal type of pattern, but due to circuit factors the actual pattern appears as in Fig. 7-25.

With such a vectorgram display, color (chroma) troubles are in evidence, and servicing procedures are thus expedited. When nonlinear distortion is present, some of the petals are longer than others. When the petal tops are flattened, signal overloading of the circuit is present. If there is a loss of the red chroma signal caused by the absence of an $R - Y$ input to the vectorscope, a collapse of the petal formation display occurs, and only a horizontal line appears. Thus, the failure of the $R - Y$ circuit is immediately evident. Similarly, if the $B - Y$ circuit is defective and does not produce an input, a vertical line appears since the horizontal sweep formed by the $B - Y$ signal is missing. Again the particular stage that is at fault has been pinpointed.

If the petals appear wider than usual and tend to overlap, the alignment of the chroma, the IF, or the bandpass circuits should be checked using a sweep generator and marker. If the petals have a blurry appearance, it may indicate the necessity for a critical adjustment of the 3.58-MHz trap. Numerous other tests can be made with the vectorscope, and a full listing of its capabilities and advanced application techniques will be found in the literature accompanying the device.

7-9. SPECIAL METER SCALES

Earlier (in Fig. 7-2) the volt-ohmmeter scale included a decibel (dB) range. This scale is useful where a comparison must be made between two levels of voltage, current, or power. An audio amplifier, for instance, can be rated as having a 3-db level above another audio amplifier. This would indicate that the power output from the first is twice that of the second. The decibel scale is for measuring *differences* in voltages, currents, or power, rather than amplitudes.

If we double the power output of an amplifier, we have changed the power by 3 dB. The latter is a valid expression regardless of what the true power was originally. Thus, if we converted a 20-W amplifier to a 40-W unit, the power increase in decibels is 3. If, on the other hand, a transmitting station of 200 kW would increase their power by 3 dB, it would mean a doubling of their power to 400 kW. Most VOM units include a decibel (dB) scale so that relative powers can be read. The bel was named after Alexander Graham Bell, the famed American inventor. The unit bel is not used. Instead, one-tenth bel (decibel) is used and represents a difference in sound intensity that is barely noticeable to the average listener.

The decibel can also have a negative value. If power is halved, for instance, it represents a −3-dB change. Decibels are also additives; that is, if we increase a 5-W power to 10 W, there is a 3-dB change. If we now go from 10 to 20 W, there is another 3-dB change. Thus, the total change from 5 to 20 W is 3 dB + 3 dB for a 6-dB difference. Because of the relationships between voltage, current, and power, a doubling of current equals 6 dB. Similarly, a doubling of voltage also equals 6 dB. Thus, a change of current from 2 to 4 A is a 6-dB change. Similarly, a voltage change from 50 to 100 V constitutes a 6-dB change. Sometimes certain reference levels are used for establishing a basis for values that are given. Thus, for instance, a 0-dB reference may be indicated as 0.006 W (6 mW) across a 300-Ω resistance.

A typical dB scale on a voltmeter dial is that shown in Fig. 7-27(a). Here the range is from −20 to +2 dB. The scale may be related directly to the ac scale, and many meters have the reference as 0 dB = 1 mW of power across a 600-Ω resistance. This could then represent 0.7746 V ac (rms) across the 600-Ω load resistance. The abbreviation dBm is used to indicate a decibel reference based on 1 mW = 0 dB at 600 Ω.

On some meters the reference point may be 0 dB = 6 mW at 500 Ω, with the 0-dB point on the meter scale at 1.732 V. Other manufacturers

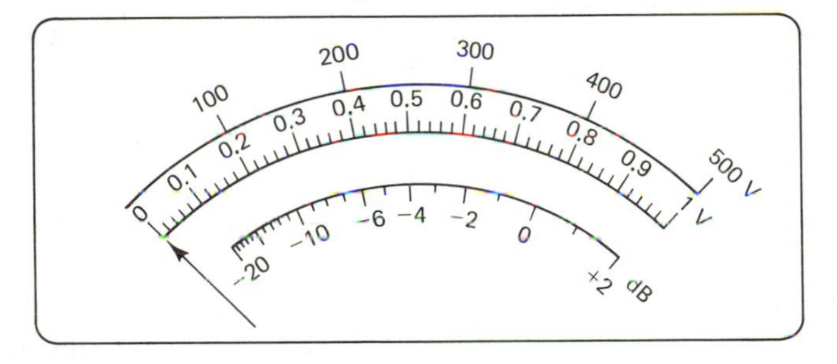

(a)

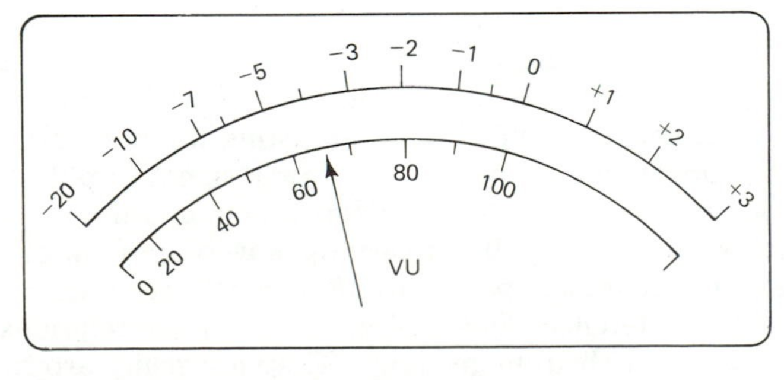

(b)

Figure 7-27 dB and VU meter scales.

may select 0 dB = 6 mW for 600 Ω at 1.897 V. For all such instances, however, dB measurements provide a direct reading on the decibel scale when the meter is placed across a load resistance having a value as specified in the instruction booklet accompanying the meter.

The decibel scale is particularly useful for comparing the power output from each stereo power amplifier channel. Thus, for a fixed voltage ac signal input the power output in decibels is measured from the right channel and from the left channel. In a well-designed stereo system the two channels should be very close in power output for a given signal input. Thus, the decibel reading for each should be virtually the same. If there is more than a 1-dB difference in the output of one channel compared to the other, check the weaker one for defective resistors or leaky capacitors. Also check voltages on each channel to localize voltage losses.

Another meter scale that can be considered a special type is the one often used as a volume unit (VU) indicator in audio tape recording and on occasion in FM/AM receivers. The VU meter dial measures sound intensity, and a typical scale is shown in Fig. 7-27(b). This scale is not usually found on a VOM or other similar test equipment since the sound intensity can be ascertained with the decibel scale. These meters are actually decibel oriented in which the zero level is assumed to equal 0.001 W across an impedance (Z) of 600 Ω. (An impedance is the combined opposition to current flow offered by a resistance combined with either a capacitance or an inductance.)

For the dial face illustrated in Fig. 7-27(b), the upper scale extends from +5 at the right to −20 at the left. The second scale directly below the first is marked from 0 to 100 and represents the *percentage of voltage*. As shown, the 100% point is directly below the zero point of the upper scale. In many meters such as this the scale markings to the right of the zero point are in red with those to the left in black.

The normal level of volume is designated as the zero VU point equal to 100% shown on the second scale. The point is usually at approximately 70% of full-scale deflection. The VU meter's maximum usefulness occurs when it is observed for short periods of time. This viewing procedure permits observation of several signal peaks, and hence the level can be regulated by adjusting the controls as needed. The volume control should be set for normal signal amplitude limits to avoid overload. In contrast to the dBm meter mentioned earlier, VU measurements relate to signals of varying waveshapes, amplitudes, and frequencies. Instantaneous peaks of such signals often extend beyond the 1-mW dBm reference. Essentially, the VU scale relationships have no fixed value as with dBm readings. They are really arbitrary values established for the primary purpose of indicating and preventing overload conditions.

8

Power Supply Testing and Repair

8-1. THE BASIC SYSTEM

The purpose for a power supply in electronic equipment is to supply the necessary direct current (dc) to operate the electronic circuits made up of transistors and integrated circuits. If a device operates on batteries, the power source is then already dc, and nothing else is needed. When, however, a unit must derive its power from the ac mains, the alternating current therein must be converted to direct current (see Sec. 1-1, Fig. 1-2). Because the ac for normal services in the home is rated at 120 V, it is necessary to step the voltage down to a lower value as required by the electronic circuitry involved. While this voltage reduction can be accomplished by series resistors, overheating and current-carrying capacities present problems. Hence, the most simple procedure is to use a step-down transformer, designated as T_1 in Fig. 8-1. Here the primary (P) is fed the 120 V from the ac mains, utilizing the on-off switch (SW) shown. The secondary (S) in this instance supplies 30 V of ac which is converted to a form of dc by a power supply diode (D_1). The diode permits current to flow only in one direction and hence produces a periodic pulse system, as shown. [See also Fig. 1-2(b).] The diode rectifier is followed by a filter capacitor (C_1) which tends to oppose a change of current and hence smooths the ripples of ac into a fairly steady dc. If this is followed by another capacitor (C_2), greater filtering is established, and consequently an output voltage is obtained having positive and negative polarities. Resistor R_1 is in series with the current flow and to some degree regulates the amount of voltage obtained at

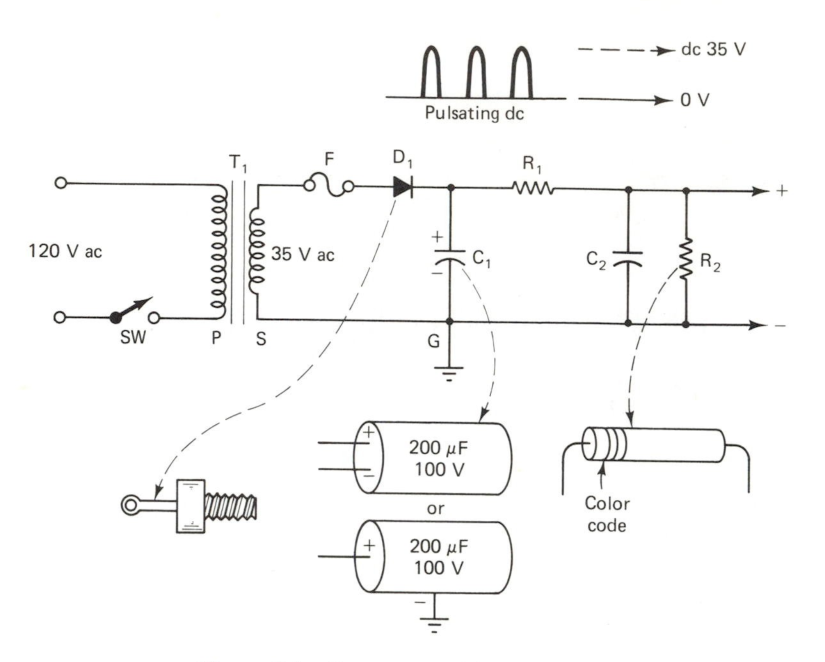

Figure 8-1 Component identification.

the output. Resistor R_2 across the output tends to stabilize the voltage output since it creates a small but constant drain on the power supply.

The power supply illustrated in Fig. 8-1 is known as a *half-wave* type since it only recovers every other alternation of the ac waveform. A better system is the full-wave type, illustrated later. For the circuit shown in Fig. 8-1, the most common troubles occur with the filter capacitors C_1 and C_2 and the diode D_1. If the filter capacitors become defective, some 60-Hz hum can develop at the output and cause interference signals in the video and sound reproduction. If diode D_1 becomes defective and produces an open circuit, there will be an absence of voltage output, and the VCR becomes totally inoperable. If diode D_1 shorts, the fuse (F) will open and again cause an inoperative VCR. Similarly, if the capacitors develop internal short circuits, it will result in excessive current drain, and again the fuse will open. If the fuse is not of proper rating, it may cause a burnout of the diode before the fuse opens. Hence, if the fuse is open, the components should be checked for shorted conditions. It is possible for a short to develop in the transformer secondary (S) and cause a fuse to blow out. This is unlikely unless an overload remains on the secondary for some time before the fuse opens. If the secondary is damaged, a new transformer will have to be installed.

8-2. REPLACEMENT PARTS

The components that most often cause trouble in power supply systems are the capacitors and rectifier diodes. If capacitors C_1 and C_2 in Fig. 8-1 become defective, they will either short out internally or develop an open circuit. For the latter condition it is as though the capacitor had been removed, and consequently some hum filtering is lost. Consequently, until the capacitor is replaced, some hum interference will appear in the video and sound reproduction of the VCR. If a capacitor shorts out, it has virtually zero resistance and hence will pass an excessive amount of current. If the proper fuse is in the circuit, the latter will open and thus shut off the power. When a fuse has burned out, it is always necessary to check the circuits to find the cause before replacing the fuse. If a fuse has a higher rating, it may not open when parts become defective, and excessive current will flow through diode D_1 as well as the transformer secondary. Hence, the diode and transformer can be damaged if the proper fuse is not present to safeguard the system.

It is unlikely that resistors R_1 and R_2 will cause any trouble unless excessive currents flow through them. A shorted C_2, for instance, will cause an excessive current flow which must go through R_1 and hence will overheat this resistor and possibly burn it out. If either resistor burns out, it creates an open circuit. An open circuit for R_1 will make the power supply inoperative since no voltage will appear at the output. If resistor R_2 opens, however, an output voltage is still obtained, though it will be slightly higher than normal because this resistor no longer provides a constant current drain for stabilization purposes.

Failure of the power supply can also be caused by an overload condition in the circuits that are fed by the supply. Thus, if a blown fuse is present, the components of the power supply should be checked initially. If these are all right, the components in the circuits *beyond* the power supply must be checked to isolate the part that is defective, as described more fully in Chapter 10. For the power supply shown in Fig. 8-1, the usual part replacement procedures prevail. Replacement components should have the same (or better) rating than those found defective. Thus, a 2-A, 50-V rectifier replacement should have at least the same rating or one that has a higher current or voltage rating (or both). A replacement such as a 3-A, 75-V type would last longer and run cooler. The disadvantages are the additional cost and the necessity for having more space to accommodate the larger size.

Capacitors C_1 and C_2 are known as *electrolytic* types and are polarized with identifying positive and negative markings. The markings may be as shown below the power supply in Fig. 8-1 with either two terminals available (positive and negative) or a single terminal (positive) with the metal casing placed at chassis ground. The latter procedure is representative of the minus polarity connection. The ground symbol (G) usually represents a common terminal using the metal chassis. It also designates a common connection on a printed-circuit board. If a capacitor is rated at 200 μF at 100 V, it should be replaced with the same type or one with a higher rating for greater filtering and voltage handling capabilities. Again, however, the larger units cost more and take up more room.

When a fuse blows out, it should be replaced only with the exact type recommended. If the rating of the fuse is too low, the fuse will burn out too often. If a higher rating than normal is used, the fuse will fail to open the circuit for excessive loads and hence will damage the power supply system. If the fuse is the correct one but blows out only on occasion, it indicates a borderline overload that creates a current flow near that of the fuse's rating. The least fluctuation will burn out the fuse. When this condition occurs, the electrolyte capacitors should be checked to see if they are becoming leaky and shorting out some current. If the capacitors are leaky, they may be warm to the touch and should be replaced.

8-3. FULL-WAVE SUPPLIES

A better power supply system than the half-wave type discussed in Sec. 8-1 is the *full wave*, illustrated in Fig. 8-2. The power supply shown in Fig. 8-2(a) utilizes two diodes (D_1 and D_2). As the successive positive and negative alternations of the ac appear across the transformer secondary winding, diodes D_1 and D_2 conduct alternately. Thus, when the upper section of the secondary winding is positive, diode D_1 conducts, and when the lower section is positive, diode D_2 conducts. In this fashion successive alternations are utilized, and the resultant unidirectional pulsations can be more easily filtered for a smoother ripple-free dc output. For this particular system, however, the transformer secondary must have a center tap (CT), as shown in Fig. 8-2(a). The voltage present on each side of the secondary coil is approximately equal to that prevailing across the total secondary of the half-wave supply shown in Fig. 8-1. Thus, if the latter had 35 V at the secondary, the one in Fig. 8-2(a) would require 35 V across each half of the transformer for a total secondary voltage of 70 V. Since only one-half of the secondary is utilized at any instant, the output voltage is approximately that obtained for the half-wave type.

The filter section at the output from the diodes is identical to that shown earlier for the half-wave type. The same replacement-parts fac-

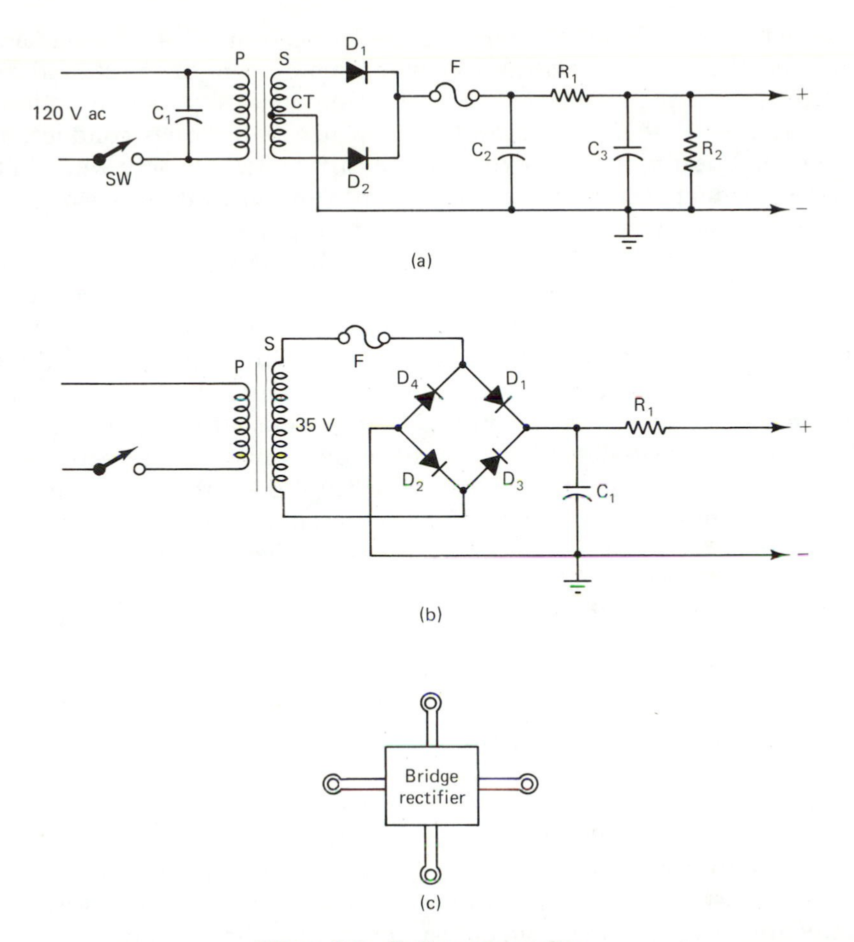

Figure 8-2 Full-wave supplies.

tors discussed in Sec. 8-2 are also applicable here. Because full-wave systems need less hum filtering, they can use lower-value capacitors. If a particular half-wave supply utilizes 500-μF capacitors, for instance, the full-wave type may provide as good (or better) filtering with 250-μF types. (The location of the fuse may vary for different power supplies, depending on the manufacturer's design.) If the diodes short out, there would be no protection against a burnout of the secondary winding. Hence, some manufacturers may have a fuse at both the top and bottom of the secondary winding. In some instances, a fuse may also be present in series with the primary winding for added protection.

Another full-wave system is shown in Fig. 8-2(b). This system eliminates the center-tapped transformer and uses four diodes instead. Such a power supply is sometimes termed a *bridge type* because of the configuration of the diode sections. In such a supply diodes D_1 and D_2

conduct simultaneously when the upper section of the secondary is positive. For the next alternation, when the bottom of the secondary is positive, diodes D_3 and D_4 conduct simultaneously. The direction of current flow is identical regardless of which two diodes conduct. The same condition prevails for the dual-rectifier power supply wherein the diodes conducted alternately but current flow was still in one direction to provide for a positive and negative output polarity.

Although not shown in Fig. 8-2(b), the same components are utilized for filtering purposes as for the earlier power supply. Consequently, the same replacement factors prevail as discussed in Sec. 8-2. Individual diodes can be wired into the system as shown in Fig. 8-2(b), though often the four diodes are mounted in one integrated housing with four external terminals, as shown in Fig. 8-2(c). Thus, even though only one diode may become defective, the entire unit must be replaced with an identical unit to restore operation. For the bridge rectifier system the common ground connection exists between diodes D_2 and D_4. The transformer secondary is across the diode bridge, and hence none of it is at true ground potential. Thus, unless the system is carefully designed, the residual hum level may be somewhat higher than with the full-wave system illustrated in Fig. 8-2(a).

8-4. COMPONENT LOCALIZATION

As mentioned earlier, a blown fuse indicates an overload condition that must be corrected before fuse replacement.

A useful tool for effective component isolation is the volt-ohmmeter unit discussed in Chapter 7. Assuming that the circuit to be tested is the half-wave power supply shown earlier in Fig. 8-1, a quick check of the diode rectifier can be made by using the ohmmeter scale. Whenever an ohmmeter is used for circuit tests, all power must be shut off to the VCR. In testing the diode it is advisable to remove the fuse so that an open circuit is formed on one side of the diode. This prevents the shunting effect of circuit resistances from giving false readings. Placing the test probes across the rectifier as shown in Fig. 8-3(a) should indicate a high resistance in one direction and a low resistance when the probes are reversed. If a low resistance is obtained for both positions of the probes, it indicates a shorted rectifier unit.

In testing capacitor C_1 as shown in Fig. 8-3(b), the capacitor should be discharged initially by placing a metal end of a screwdriver (or a piece of bare wire) across the terminals. Next apply the text probes across the capacitor, again using the ohmmeter setting. Since this capacitor is an electrolytic type, it will have a high reading in one direction and a low reading in the other. A low reading for both direc-

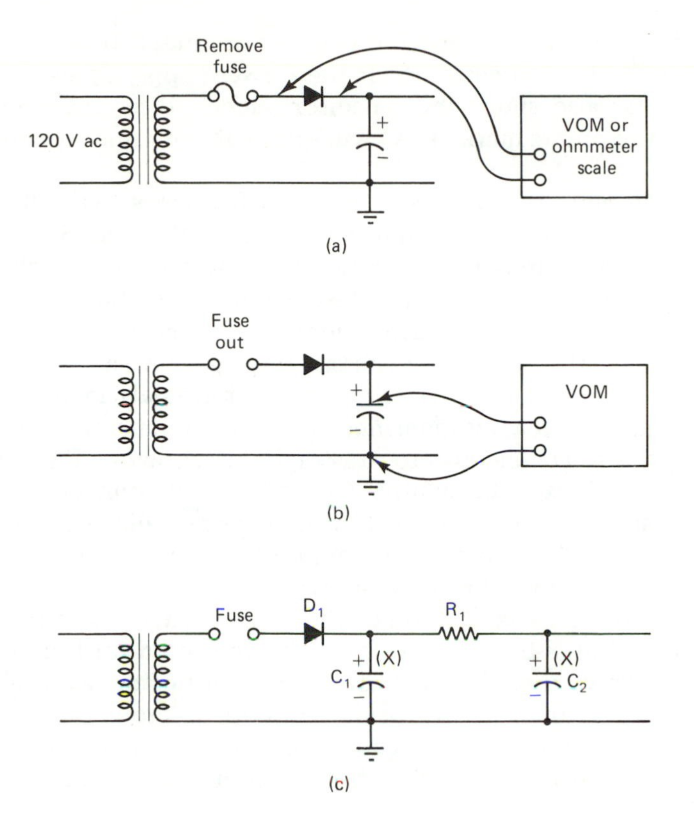

Figure 8-3 Test points on half-wave supply.

tions (when the probes are reversed) would indicate a shorted condition. A high reading in both directions indicates an open-circuit condition. In either case the capacitor must be replaced.

In testing capacitors and diodes, the meter scale should be on a medium range rather than set at a high or low extreme. At a high-range ohmmeter setting a comparatively low reading will be obtained for either direction, and such readings can be misleading. For an extremely low-ohm scale, no reading may be obtained at all. As shown earlier in Fig. 8-1, the capacitor may have two terminals or one terminal but with the metal casing acting as the other terminal. Thus, for the latter condition, one ohmmeter probe is placed either on the metal casing of the capacitor or on the metal chassis.

In checking components shunted by other units, consideration must be given to their effect on readings. Thus, for meter probes placed across C_1, we have a series network consisting of R_1 and C_2, as shown in Fig. 8-3(c). A more accurate check of either capacitor can be made by disconnecting one side of each capacitor before testing it. The best method for doing this is to disconnect the positive terminal of the capacitor but to leave the ground connection intact. It must be remembered that the load circuits which the power supply feeds (such as the VCR circuits) also shunt the capacitor. Hence, disconnecting each capacitor at the points marked (X) will eliminate any questionable effects of loading.

In examining the underside of a VCR chassis, we must learn to identify the terminals of the components to be checked. Since the chassis layout varies from one manufacturer to another, certain clues can be used to facilitate identification and isolation of specific units. As shown in Fig. 8-1, capacitors are usually identified by markings and hence can be localized on the top of the chassis. Having found a particular capacitor, the chassis can be inverted and the terminals identified. If the capacitors are not readily identified, it helps to remember that C_1 connects to rectifier D_1 and also to resistor R_1, as shown in Fig. 8-4. Having thus localized C_1 we can identify C_2 also since it connects to the other side of R_1 as well as R_2, as shown. Also, having localized D_1, the positive lead can be traced to the fuse holder and from there to the transformer mounted on the upper chassis, as shown.

An alternate parts location of the components shown in Fig. 8-4 is illustrated in Fig. 8-5. Here a single housing is used for both capacitors C_1 and C_2. Thus, there is a slight rearrangement of the associated units, as illustrated. With a single casing for two or more capacitors some space is saved, though if one capacitor becomes defective, the whole unit may have to be replaced. If there is room on the chassis, however,

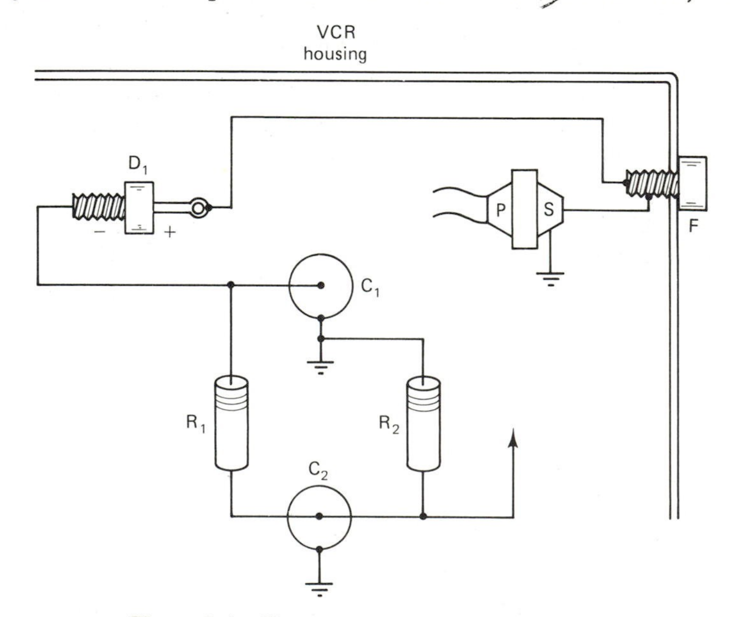

Figure 8-4 Chassis component localization.

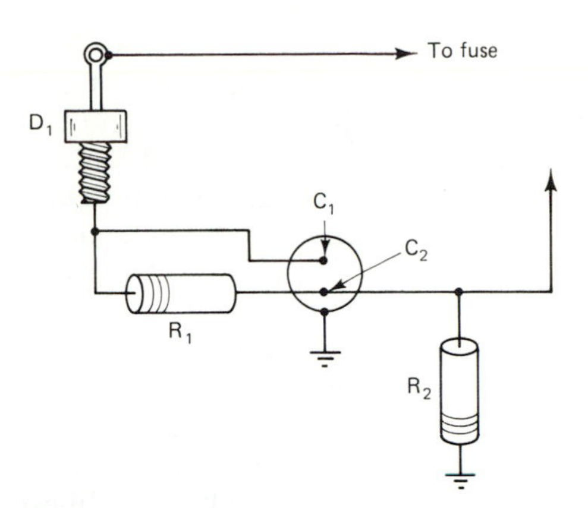

Figure 8-5 Alternate parts location.

the defective one can be disconnected and a single replacement capacitor wired to the appropriate terminals.

Details of the possible placement of the transformer primary leads are shown in Fig. 8-6(a). Here the primary coil of the transformer is fed by the incoming ac line, but one section is interrupted by the on-off switch, as shown. It is essential to trace these wires in case the transformer has to be replaced or if problems arise with the switch. In some instances an electronic switching device is used, as shown in Fig. 8-6(b). In either arrangement, an inoperative VCR requires some preliminary checking of the input power system to make sure the ac lines, switches, and transformer are not defective.

If the switch is tripped but fails to turn the VCR on, it could, of course, indicate a faulty wall socket. To check this, the VCR line cord and plug should be removed from the wall socket and the latter tested by plugging in a floor lamp. If the lamp lights, the trouble may be in

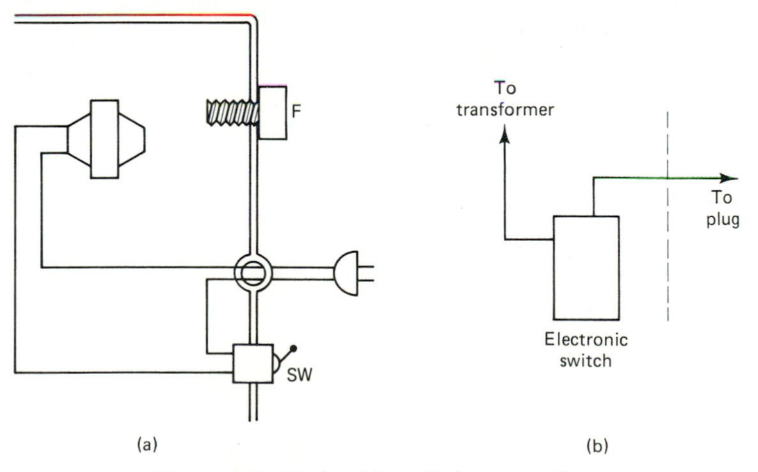

Figure 8-6 Underside switch connections.

the VCR line cord or its ac plug. For the circuit shown in Fig. 8-6(a), disconnect the plug from the ac line and place an ohmmeter across the plug terminals. Now flip the on-off switch, and if proper continuity exists through the transformer primary, a reading should be obtained on the ohmmeter scale. Do not use too high a meter scale for this check. If no reading is obtained, check the line running to the switch by placing one test probe to a plug terminal and the other to a switch terminal, as shown in Fig. 8-7.

The test points are marked with alphabetical symbols for convenience in describing the procedures involved. If no reading is obtained between points (A) and (B), it indicates an open-circuit condition. The latter may be caused either by an open circuit in the connecting wire or a defective plug. If a reading is obtained, the probe should be removed from (B) and placed at (D). Now with the switch in one position a reading should be obtained, but an open circuit should exist when the switch is thrown in the other position. If an open circuit exists for either switch position, it indicates the switch is defective and must be replaced. If the switch proves to be in working order but no power is present in the VCR, one probe should be placed at (C) and the other at (E) to check the unbroken line. If no bare terminal is available at (E), some of the insulation must be cut away to gain access to the wire. Again, if no reading is obtained, the ac line cord is defective and must be replaced.

If a reading is obtained, a double check can be made between points (E) and (D). A reading should be obtained, indicating continuity in the transformer primary. If no reading is obtained, the transformer is defective and must be replaced. If continuity is obtained, the meter can be placed on the ac range for reading the 120-V line voltage. The plug is then placed in the wall socket and a reading taken across (E) and (D) of Fig. 8-7 to check for the presence of line voltage. Flip the switch several times to make sure it operates properly. If voltage is present but the VCR is inoperative, check for voltage across the secondary at ter-

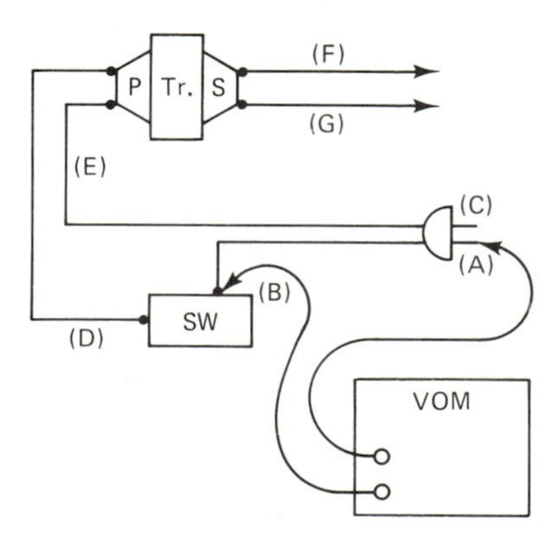

Figure 8-7 Switch system check points.

minals (F) and (G). If no voltage is obtained here but voltage is present across the primary, the transformer is defective and must be replaced.

8-5. FULL-WAVE TESTS

For the full-wave dual-diode rectifier shown earlier in Fig. 8-2(a), the localization and testing of resistors and capacitors follow along the same lines detailed in Sec. 8-4. Since two diode rectifiers as well as a center-tap transformer are involved, some test variations are required. Because the diodes have a common connection as shown in Fig. 8-8(a), false readings may be obtained with in-circuit component testing. An open fuse does not affect the closed circuit formed by the secondary winding in series with the two rectifier diodes. However, if diode D_1 is defective (shorted), an ohmmeter reading across this diode will disclose the condition. If diode D_2 were not shorted, however, a reading across it would show a virtual short because of the low resistance of the transformer secondary plus the shorted condition of D_1. Consequently, it is preferable to disconnect one side of each diode to get an accurate test that would show a high resistance in one direction and a very low resistance in the other direction. A convenient point for opening the

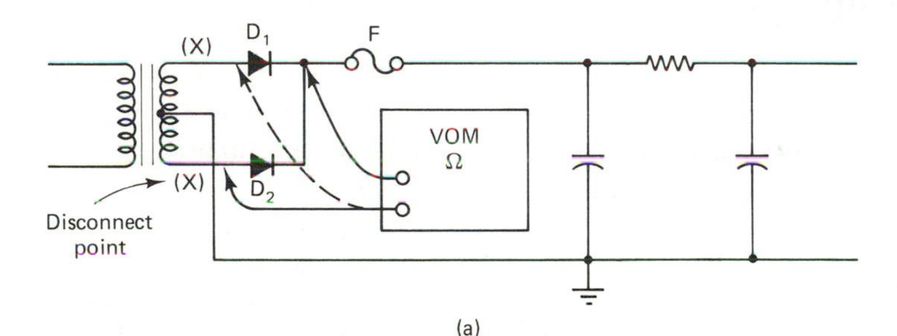

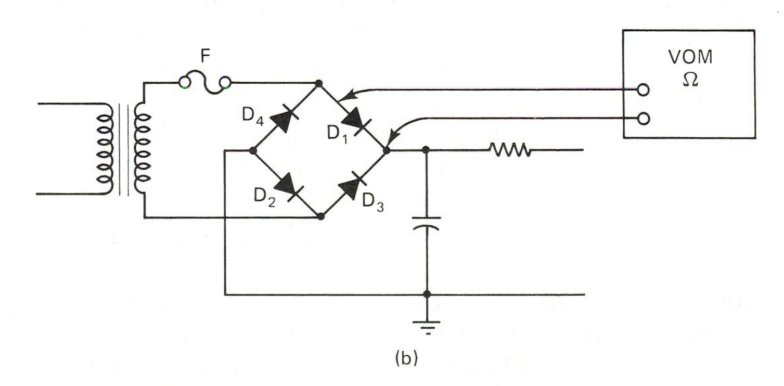

Figure 8-8 Full-wave system check points.

circuit would be at the positive side identified by (X) in Fig. 8-8(a). If a fuse is on the secondary side of the diodes, it can be removed to form an open circuit for test purposes.

For the bridge network shown in Fig. 8-8(b), a quick check can be made by placing ohmmeter probes across a diode, as shown. Since other diodes (plus the secondary) are also involved in the circuit linkages, it is again advisable to disconnect a diode on one side so that reliable readings can be obtained. Since the bridge network is a balanced circuit, a defective diode should be replaced by one having characteristics very similar to the others. If a defective diode is replaced by one having higher voltage and current ratings, there would be a difference in the internal resistance. Hence, during the nonconducting state the bridge circuit balance would be disturbed.

Note that an open fuse does not remove shunting units in a bridge rectifier system. With an open fuse in the circuit at Fig. 8-8(b), any particular diode is still shunted by three others. In addition, the circuits in the VCR are also in shunt across the power supply and hence the diodes. The best check is still formed by opening one side of the line connected to the diode under test.

8-6. OVERLOAD SYMPTOMS

In all electronic equipment overload conditions produce certain symptoms that help localize the defective parts. Under normal circumstances a well-designed VCR is protected by adequate fusing, as more fully illustrated in Sec. 8-7. Thus, if a partial short develops in components and circuits, a fuse will open. Fuses are designed to open the circuit and thus shut off power at the threshhold of overload. In essence the fuse senses the danger that could result from an overload on the circuits and thus protects them. If, at some time or another, a fuse has been replaced by one having a higher rating than the original, the danger of an overload causing equipment damage is increased. If, for instance, a 3-A fuse was part of the original equipment and was subsequently replaced by a 5-A fuse, the VCR would operate normally with the larger fuse so long as no circuit problem occurs. The VCR circuitry would be liable to be damaged when an overload causes current drains between 3 and 5 A. Some circuits and components may be damaged quickly for a small overload, while others may take a short interval of time before being damaged.

One basic symptom of an overload condition is the buildup of heat. Electrolytic capacitors, for instance, should not be hot to the touch. Any internal heat buildup indicates a short or a partial short that causes an excessive amount of current flow. Thus, such capacitors should be replaced immediately. At the same time the fuse associated with the capacitor circuitry should be checked and replaced if it exceeds the recommended rating. Transformers may also exhibit excessive heat, though often power transformers are warm or even slightly hot to the touch, depending on design. If, however, the overheating of the transformer causes a scorching of the coil insulation, a strong odor will be evident. Under these conditions the cause of the overload would have to be determined (shorted capacitors or diodes) and replaced. After the transformer has cooled down, the unit can then be operated again to see if the transformer has not been damaged.

Some components operate at fairly high temperatures under normal conditions. Rectifier diodes, for instance, may become quite hot and often are mounted on heat-dissipating metal flanges to keep temperatures below the point where damage will occur. Some resistors may also become quite hot, particularly the wire-wound types that carry 0.5 A or more of current. A few composition-type resistors may run quite warm under normal operating conditions. If the VCR is operating satisfactorily and the fuse has the proper rating, such resistors are operating satisfactorily and need not be replaced.

8-7. FUNCTIONAL VCR SUPPLY

Because design considerations vary from one manufacturer to another, the power supply systems for the various VCR units on the market vary to some extent. However, since the general operational principles and requirements are the same for all VCRs, certain basic sections are needed in each. A representative example of a complete power supply is shown in Fig. 8-9. Here the bridge rectifier circuitry discussed in Sec. 8-3 is used. Variations in the overall circuitry may consist of different output voltages, more critical voltage regulating systems, and a greater degree of filtering if needed. The essential sections, however, are represented in Fig. 8-9.

The ac-line input plug is shown at the upper left of Fig. 8-9. Fuse F_1 is one of the several utilized for protective purposes. This fuse protects the line system from a shorted C_1 capacitor or from a short circuit developing in the primary or secondary of the transformer. The inductors L_1 and L_2 in conjunction with capacitor C_1 form a noise-filter network that isolates to some degree the power supply from noises present in the ac input line. The L_1 and L_2 coils in series with the input to transformer primary P_1 act as a trap to keep out high-frequency noises that may enter from the ac line. If these noises ride through the power supply circuit, they can reach the sensitive picture and sound circuitry and cause interference. Similarly, capacitor C_1 acts as a shunt filter to

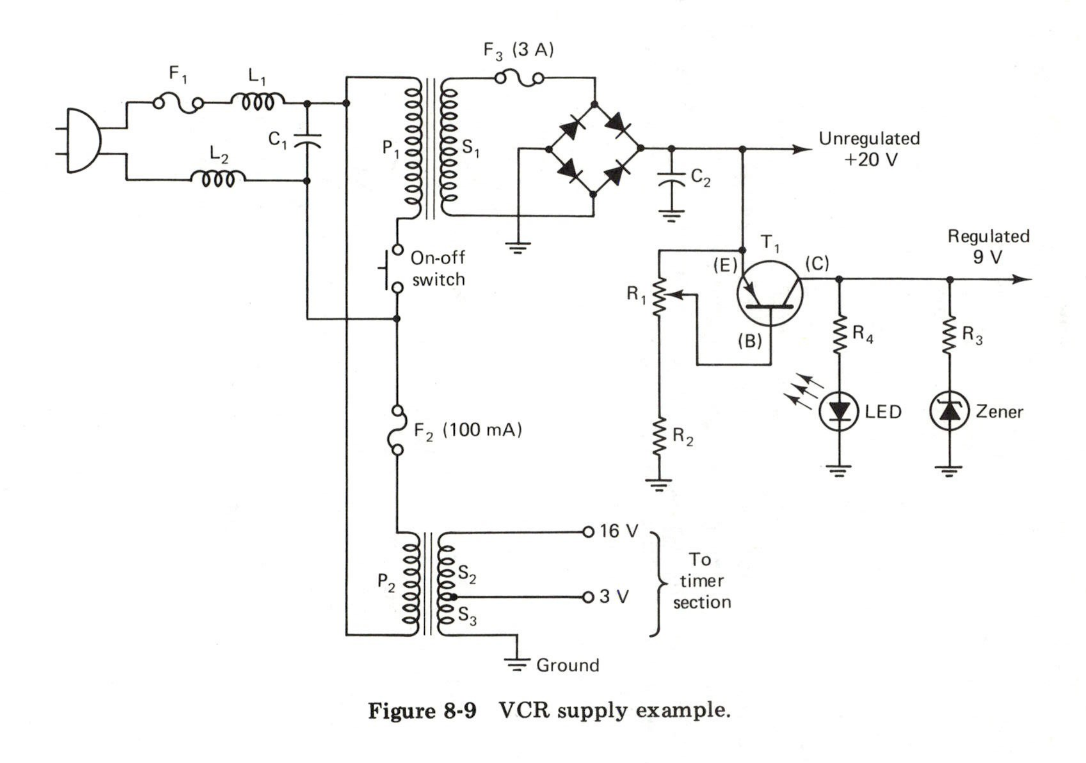

Figure 8-9 VCR supply example.

bypass any high-frequency noises that were not entirely eliminated by the series coils.

Note that the incoming ac line feeds both the primary of the upper transformer as well as the primary of the lower transformer. The on-off switch, however, is only effective for the upper primary (P_1). Thus, when the ac plug is placed into the wall socket, power is applied to transformer primary P_2 whether or not the on-off switch is in the open or closed position. This arrangement is necessary to keep the timer section of the VCR in constant operation, even though the unit may not be in use for recording or playback. The timer section contains an electric clock, control knob, and switches for automatically recording programs at some future time. As shown, a separate fuse (F_2) is in series with primary P_2 for protection of the timer section in case of overload conditions. Note that the secondary section of the lower transformer is divided into two parts S_2 and S_3. Thus, two voltages are available as needed.

The upper transformer secondary (S_1) is applied to a bridge rectifier circuit of the type discussed in Sec. 8-3. Here a 3-A fuse (F_3) protects the secondary coil against any overload conditions that may occur. The output line from the bridge section is bypassed by filter capacitor C_2 before it is applied to the VCR electronic circuitry. The output for this particular section is an unregulated 20 V under normal operating conditions. A tap on this 20-V output line is applied to transistor T_1. The latter, with its associated circuitry, decreases the voltage it obtains to a 9-V output, as shown. To maintain this voltage as close to the required amount as possible, a *zener* diode is utilized. The zener diode has the ability to maintain a constant output voltage within certain limits regardless of slight variations in the amount of current drawn by the VCR.

Resistor R_3 develops a voltage drop so that the zener operates within a proper range. Resistor R_1 (between the emitter and base terminals of the transistor) is set for the optimum output voltage from the collector terminal of the transistor. This permits the zener voltage-regulating diode to operate within its normal voltage range. If this diode must be replaced, care must be taken to observe the polarity, as shown. This is necessary because the zener operates in its negative region and will not function properly if the terminals are reversed.

Resistor R_2 completes the transistor input circuitry by providing a ground-return circuit. If any of these resistors must be replaced, be sure to use exact values of resistance and wattage ratings. A light-emitting diode (LED) is across the transistor output terminal, as shown. Resistor R_4 provides the necessary voltage drop to the proper value required by the LED. The latter furnishes the indicator light when power is applied to the bridge rectifier circuit and hence to the electronic circuitry system of the VCR, exclusive of the timer section.

8-8. CHECKING FUNCTIONAL SUPPLY

Factors regarding replacement parts, component localization, and full-wave tests were covered earlier and also apply to the system shown in Fig. 8-9. Additional tests are determined by the nature of the fault and the components present. If the LED does not light up, the voltage should be measured from the top of resistor R_4 to ground to ascertain whether or not the required voltage exists. If the normal voltage is attained, the voltage should be read from the bottom of R_4 to ground to check whether or not resistor R_4 has opened. If voltage is present but the LED does not light up, it must be replaced. Absence of output voltage would also affect the operations of the electronic circuitry fed by the 9-V output from T_1. Lack of voltage may be caused by a defective transistor, misadjusted setting of R_1, or a defective R_1 or R_2. If resistors check out satisfactorily and the input voltage is present at the emitter of T_1, the lack of output voltage is probably caused by a defective transistor. A reduction in the unregulated +20-V output may be caused by a partially shorted C_2 or an overload in the system to which the 20-V output line is attached. For any substantial overload, however, fuse F_3 should open. (See also Sec. 8-6.)

As mentioned earlier, if fuse F_2 opens, it will indicate that more than 100 mA of current is being drawn. Thus, the timer section should be checked for defective components that may be causing a drain on this transformer. If either F_1 or F_2 opens, it could mean short circuits in the winding of the secondary coils S_1, S_2, and S_3. Generally, however, transformer secondary windings are of heavier copper wire than the primaries, and the likelihood of their burning out is rare. If, of course, the fuse does not protect the secondary, an excessive amount of current is drawn through the wire during the overload. The consequence is an overheating, an insulation burnout, and a short-circuit condition between the turns of wire. In such an instance the transformer must be replaced to restore proper operation. Whenever a fuse or a transformer burns out, the cause should be investigated and corrected before transformer replacement is undertaken to avoid damaging the replacement unit.

8-9. HIGH-VOLTAGE SYSTEM

A special power supply system is the high-voltage type not found in VCRs but utilized in VCR monitors and television receivers. The purpose for the high voltage is to supply a high beam-accelerating potential to the inner elements of a picture tube. The high voltage (positive in polarity) attracts the beam of electrons (negative charge). The high-voltage systems have low-current capabilities for safety purposes. Precautions are necessary, however, because the high voltages range between 25 and 30 kV and present some shock hazards, as detailed in Sec. 8-10.

The high-voltage system in monitors and receivers is actually an extension of the horizontal sweep frequency circuitry. The basic sections of the full system are shown in Fig. 8-10. Here an oscillator circuit generates a signal having a frequency locked in by the incoming horizontal sync pulses. Since the latter are broadcast with the television signals, the entire system is a remote control device. The sync pulses are transmitted by the broadcast station and used in the receivers to lock in the vertical and horizontal sweep signals formed within the receiver.

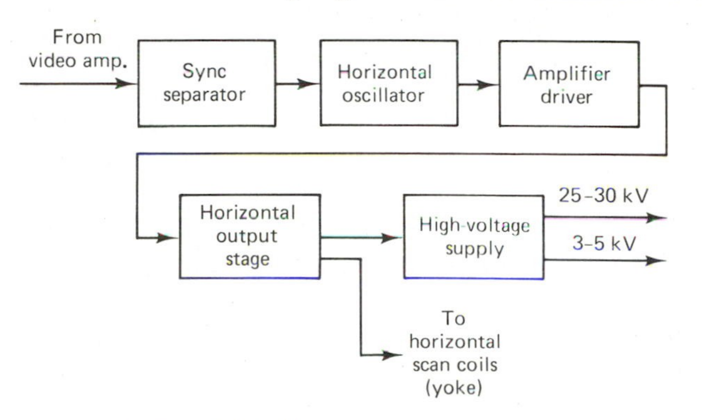

Figure 8-10 Horizontal scan system.

For the horizontal system shown in Fig. 8-10, the output from the horizontal oscillator is applied to an amplifier-driver circuit, which, in turn, supplies the horizontal input signals to the horizontal output amplifier, as shown in schematic form in Fig. 8-11. Here the output from the horizontal amplifier feeds a section of an autotransformer (single winding). The output consists of the horizontal sweep signal that is fed to the horizontal coils around the neck of the picture tube. The latter cause the beam to travel across the tube face to trace out one horizontal line at a time. Thus the horizontal sweep function requirements for the procedure are met, and the picture tube beam scans across the tube face to trace out the video images. A portion of the horizontal output amplifier signal consists of high-amplitude pulse waveforms having a repetition rate of 15,734 Hz (for color). Since these pulses have periodic amplitude changes, they behave as an ac waveform and hence can be transferred from a transformer primary to a secondary winding.

The secondary winding at the top steps up the voltage amplitude

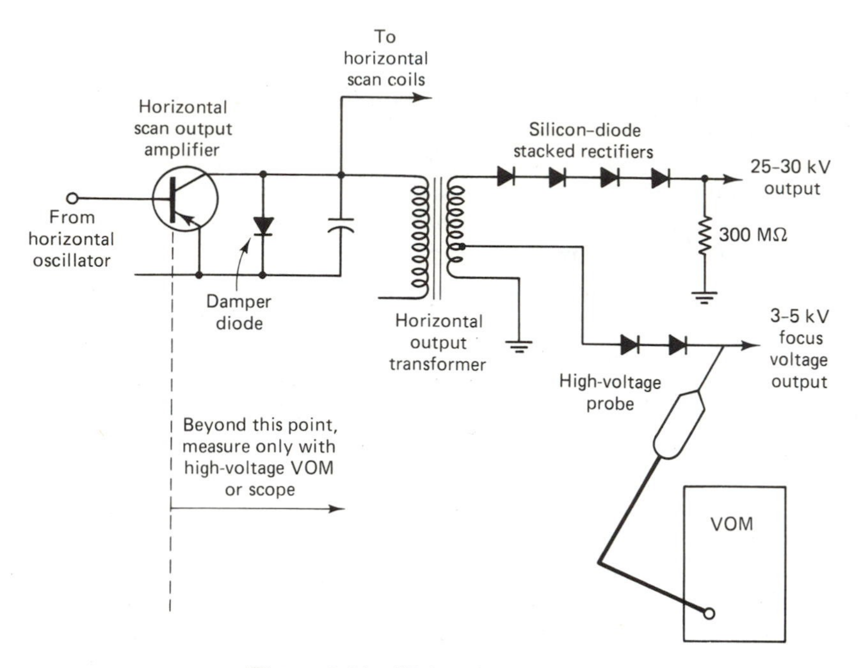

Figure 8-11 High-voltage system.

to that required by the picture tube. As shown, stacked silicon high-voltage diodes in series form the rectifier section that converts the ac-type signal to dc. A tap on the secondary winding provides for a lower-voltage output (approximately 5 kV) for the focus electrode, as detailed more fully in Sec. 8-10. A high-voltage regulating system is also utilized for maintaining the potential at a fairly constant amplitude.

The damper diode in the collector of the horizontal output transistor conducts for the peak amplitude signal peaks and hence reduces transient potentials. Additional secondary coils are used to obtain pulses at the horizontal sweep rate. One of these coils supplies the keying pulse to the AGC system (see Sec. 4-9). Another output supplies the pulse required for the bandpass amplifier, as explained in Sec. 10-12.

8-10. TESTING HV SYSTEMS

If the high-voltage power supply fails to deliver the required potential, the picture brightness suffers. Total failure of the high voltage results in a blacked-out video screen. Also, with decreased high voltage the brillancy control is ineffective. When the high voltage drops below normal, the beam velocity decreases, and the picture will expand (bloom) slightly.

In testing whether or not the high-voltage system is functioning, it is necessary to use a high-voltage probe with a voltmeter, as shown in Fig. 8-11. Series resistors in the high-voltage probe drop the voltage sufficiently so it can be read on the VOM scale. Thus, if the 0–50 scale is used, a reading of 25 indicates a multiplication of 1 MV and would indicate 25 MV. No meter or scope checks should be made at the collector terminal of the horizontal output transistor. Again, a high-voltage probe must be used to reduce the signal to an amplitude that can be handled by the scope or meter.

Lack of high voltage could be caused by failure of the horizontal oscillator circuit or the amplifier stage that follows it. Thus, scope tests should be made to find the stage that is defective. Also check the voltages as with the other circuits discussed earlier. When the defective stage has been found, a check of the transistor and associated components must be made to pinpoint the defective part. Representative test points are shown in Fig. 8-12.

8-11. PICTURE TUBE CONNECTIONS

The vertical and horizontal sweep circuits of a monitor or television receiver apply signals to a vertical as well as a horizontal deflection coil. The two coils are housed in a common container termed a *yoke*. A color tube neck may also contain convergence-adjusting magnets and other selections, as illustrated in Fig. 8-13. The picture tube has a multiple-pin input, and the interconnecting wire harness from the chassis contains a socket which is plugged into the protruding pins. The internal elements are shown in Fig. 8-14, and most of these terminate at the bottom of the tube neck where protruding pins plug into the tube socket. Note that the lower voltage (approximately 5 kV) is applied to the focus electrode of the tube. The high voltage of 25 kV or higher is connected to the inner conductive coating of the picture tube. This inner coating acts as the second anode. (The first anode is a cylindrical structure within the tube and increases beam velocity as well as helping to sharpen the beam.) On the second anode the high voltage functions as a beam accelerator. The beam, formed by the electron emission from the cathodes, ends at the inner face of the picture tube (which is coated with a phosphor material that fluoresces). Dot segments of primary colors are present on the inner face plate and are used for color picture reproduction. As the beam strikes the color dots, they fluoresce.

The inner conductive coating of the picture tube and a similar outer coating form a capacitor with the glass as the dielectric. Consequently, this capacitor acts as a high-voltage filter section that smooths out the ripple component. This factor presents a danger because the capacitor will store the high-potential charge. Thus, if the bare ends of

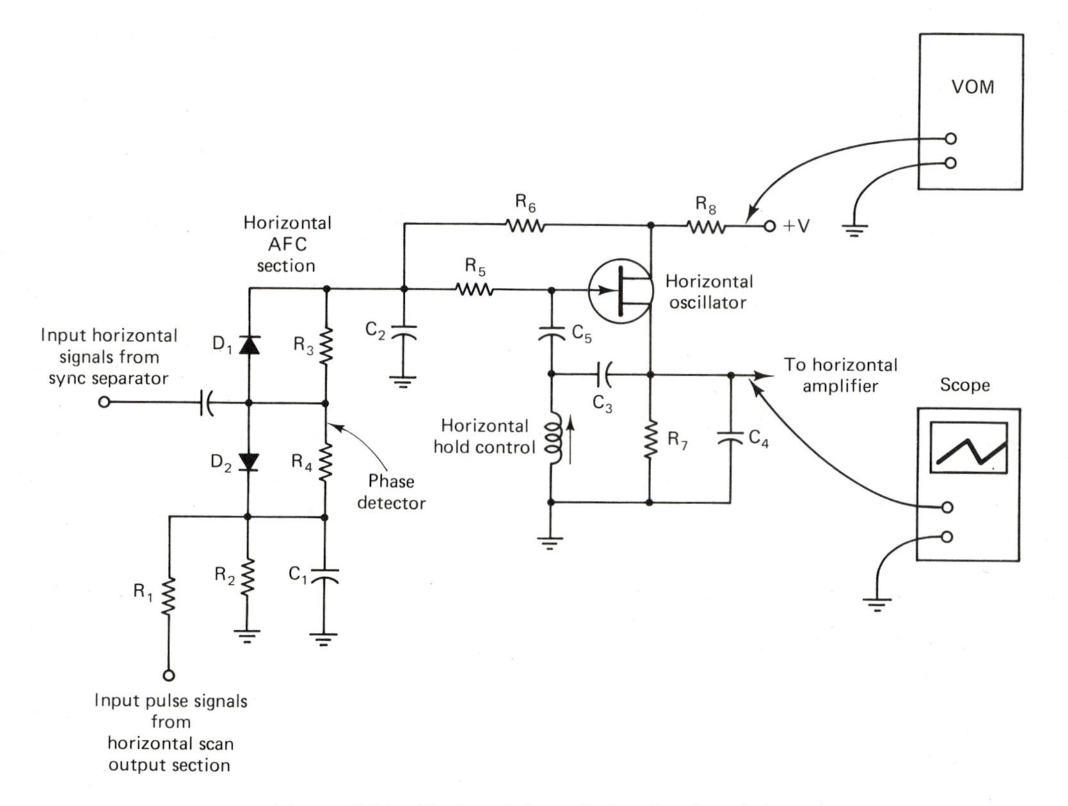

Figure 8-12 Horizontal oscillator check points.

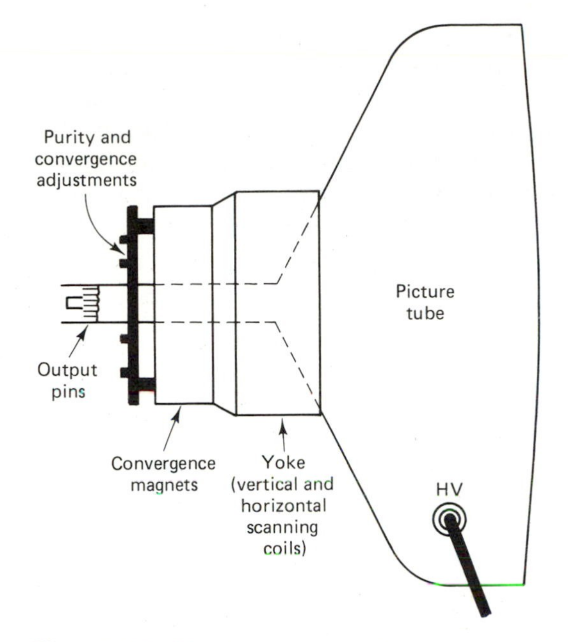

Figure 8-13 Monitor picture tube components.

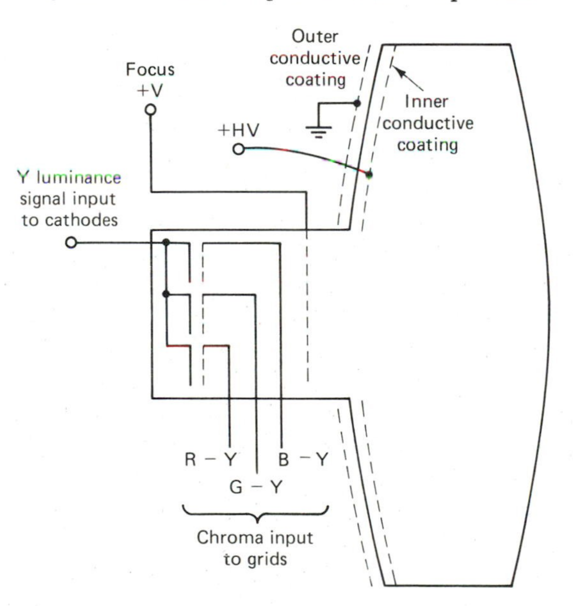

Figure 8-14 Basic tube elements.

the high-voltage feed line are touched or if the feed line is removed from the tube and the access terminal to the inner conductive coating is touched, a severe electric shock will result. The coated glass of the tube will actually hold a high-voltage charge for a considerable length of time, depending on linkage factors and atmospheric conditions. Thus, during any servicing procedures it is essential to take all precautions not only for the picture tube high-potential charge but also for the lower-voltage charges in other capacitors such as the bypass type or power supply filtering type. All these can hold a high-level charge that produces a considerable shock when touched.

9

Tuning, Aligning, and Adjustment of Sections and Circuits

9-1. FACTORS AFFECTING QUALITY

As with other electronic equipment, VCRs and monitors often exhibit trouble symptoms that can be corrected by alignment and tuning adjustments rather than by faulty component replacement. Misadjustment of controls, poor alignment, improper color control settings, and incorrect color phase and tint control setting will produce numerous symptoms of poor picture quality. Yet the latter must be recognized as not being caused by a faulty circuit or defective component. Often only slight adjustments are necessary, while on other occasions considerable retuning may be necessary in circuits not having automatic frequency facilities. If tuning and alignment procedures are ineffectual, then parts isolation factors discussed in Chapter 10 must be utilized. Reference should be made to Table 11-1 in Chapter 11 for help in locating the circuit or section that needs correction.

9-2. TUNER TRACKING

In modern tuned solid-state circuitry good frequency stability has been achieved, particularly when varactor diodes are used in conjunction with automatic frequency control systems. Most manufacturers also have excellent quality control during assembly and thus achieve good frequency stability in tuners and IF circuits. With tuners, in particular, utilization of the automatic frequency control or fine-tuning control circuitry assures automatic pull-in of the channel signal even though the tuning circuitry is not adjusted to the precise resonant frequency. In modern VCR systems, as with television receivers and monitors, the initial station selection process for one-button or touch tuning may consist of adjusting a tuning set screw to the approximate center frequency. Once all the channels that are to be received have been processed in this manner, the automatic frequency control systems take over and maintain fairly precise tuning throughout.

Similarly, the tuner tracking is set at the factory and consists of adjusting the tuning functions of the mixer, the RF amplifier, and the tuner oscillator. The tracking process thus aligns the frequency of the oscillator with that of the incoming signal so that the mixing process produces the same intermediate-frequency signal for any station tuned in. (See Sec. 1-3.) Thus, tuner tracking would entail the adjustments of the tuning set screws for the oscillator, mixer, and tuner circuits so that the proper IF frequencies are those needed for video and sound. Thus, two IF signals are involved, the video of 45.75 MHz and a sound IF of 41.25 MHz. The placement of the resonant points for the picture and sound carriers must be at those points along the response curve, as was shown earlier in Fig. 1-10(b). Thus, the 45.75-MHz picture IF is approximately halfway up the slope of the response curve. The sound IF of 41.25 MHz is at the dip in the response curve at the left, as also shown in Fig. 1-10(b). The dip is due to the trap employed to attenuate this signal and thus minimize cross interference between the video and sound signals in the receiver.

If tuner tracking is indicated in rare instances, the general equipment setup is as shown in Fig. 9-1. Here the sweep generator (with in-

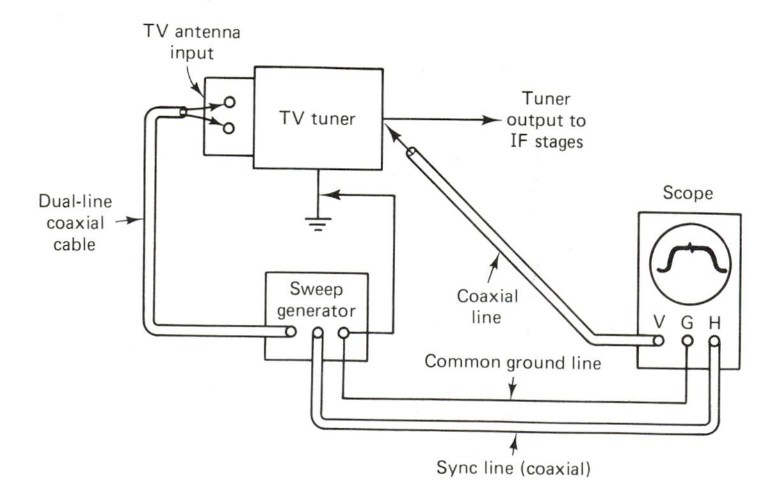

Figure 9-1 Tuner tracking linkages.

ternal marker) is applied to the antenna input terminals. (See Secs. 7-5 and 7-6.) The sweep generator is tuned to the center frequency of the IF bandpass shown earlier in Fig. 1-10(b), and the degree of sweep is set to extend slightly beyond the frequencies of the adjacent-channel picture and sound signals, as shown. The sweep generator market control is then adjusted to ascertain the correct frequency along the waveshape pattern, as shown in Fig. 9-2. If any discrepancy is evident, the appropriate tuning set screws for the tuner must be adjusted until the proper frequencies occur along the response curve. Since the location of tuning adjustments and tracking procedures may differ for VCRs and receivers of different manufacturers, it is strongly recommended that the service notes for the particular VCR or monitor be consulted for more tracking details. Since the oscilloscope is attached to the mixer output or test point, if available, it may tend to load down the tuner system. If this is the case, use a 10-kΩ isolating resistor in series with the scope lead and the mixer test point.

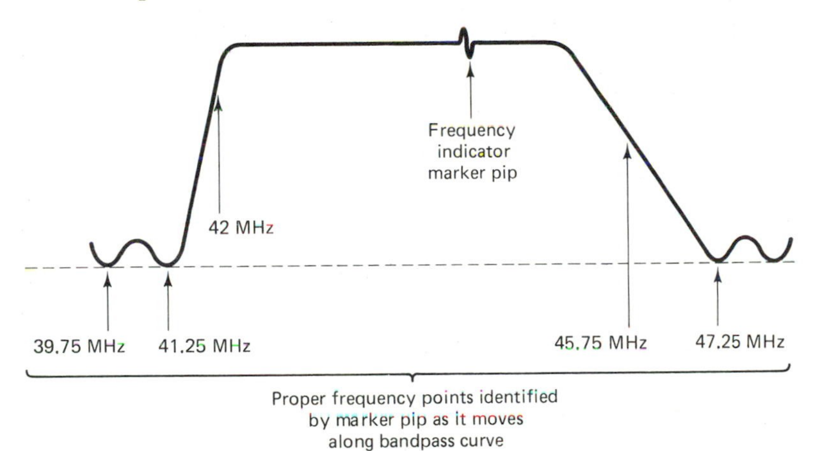

Figure 9-2 Proper frequency points on bandpass.

9-3. BASIC PRECAUTIONS

In any tracking or aligning procedures some basic precautions must be taken to obtain satisfactory results. The following list of such basic precautions is applicable to tracking and alignment procedures regardless of the type VCR or monitor involved:

1. Make sure the sweep generator and the test point on the VCR or monitor are not mismatched to the degree where poor results are obtained. Thus, the sweep generator should have a shielded output lead (coaxial cable) if the antenna input of the VCR or monitor

has coaxial cable input plugs. Do not attach a coaxial cable line from the generator to a 300-Ω input of a television receiver or other tuner.

2. Do not attempt tuner tracking unless the marker generator within the sweep generator is accurately calibrated. For the connections shown in Fig. 9-1, a fairly high output signal is necessary from the sweep generator, and the scope should have better than average sensitivity.
3. When aligning IF stages, an isolating capacitor may be required in series with the sweep generator to prevent the shunting of transistor bias voltages. The use of a capacitor across the output of the oscilloscope may be necessary to reduce spurious signal interference.
4. All test equipment should be interconnected by a common ground line, as shown in Fig. 9-3. General precautions should be observed with respect to the possibility of a "hot"chassis. Check with an ac voltmeter to see that no ac potentials can be read from the metal chassis of the VCR or monitor to a ground connection such as the outer shield covering of an incoming coaxial cable TV line or a water pipeline. Serious short circuits and danger of severe shock can be avoided by investigating the presence of high voltages on the chassis.
5. When using the sweep-sync output of the sweep generator to initiate the horizontal trace of the scope, make sure the internal sweep of the scope is turned off. Do not, however, shut off the

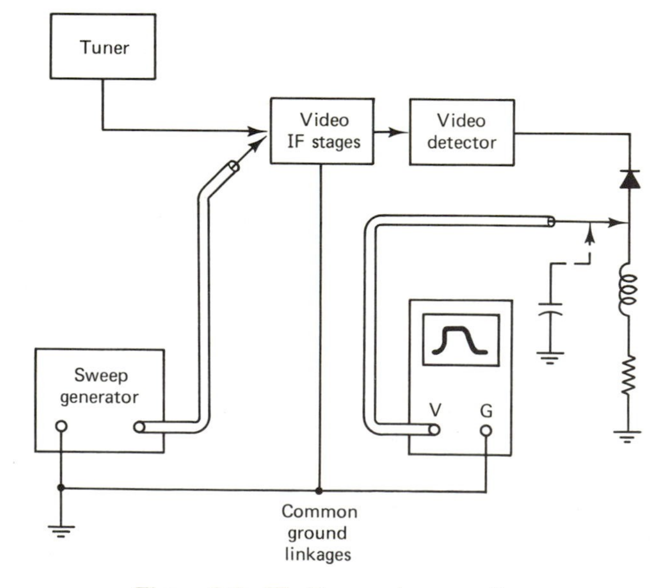

Figure 9-3 IF alignment connections.

horizontal sweep amplifier since the latter is still necessary even though the sweep signal is obtained from the sweep generator.

6. Be careful not to overload circuits by the sweep generator. If the generator signal output is excessive, it will saturate the amplifier circuits and produce a constant-amplitude output, regardless of slight tuning adjustments. Maintain the output at as low a level as possible while still obtaining a satisfactory scope image. Adjust the vertical control of the scope to increase the heighth of the scope pattern as desired.
7. Some problems may occur during initial alignment procedures if the equipment is used too soon after turning it on. The sweep generator and scope should be turned on for approximately 10 min to stabilize circuitry as well as frequency.
8. Do not overlook proper trap adjustments, even though the VCR or monitor is used in a signal area devoid of adjacent-channel reception. Improper adjustment of the traps could result in trapping out signals that are part of the desired response curve.
9. During alignment the AGC is usually disconnected since it interferes with the gain and thus gives false tracking and alignment results. To prevent the AGC circuits from giving false output readings during alignment, the AGC line should be made inoperative by opening the line that links the AGC voltages between the demodulators and the IF as well as tuner stages.
10. Improper alignment reduces the gain of the amplifier stages. Hence, as proper alignment is initiated, the gain of the stages increases. Consequently, the scope pattern on display also increases in amplitude. If the receiver is out of alignment to a considerable degree, the scope pattern would overreach the scope's tube face as alignment corrections were made. Thus, during the alignment procedures, the output from the sweep generator must be decreased. The proper procedure is to reduce the sweep generator output after each alignment step to maintain the proper level of signal display on the scope. If the IF stages are overloaded, the output will be constant, as mentioned earlier. Although this produces a flat pattern on the scope, it is an incorrect display, and the sweep generator must be reduced to a minimum for proper results.

9-4. IF ALIGNMENT

As with the tuner tracking, IF alignment should not be undertaken unless the necessity for it is clearly indicated. It is rare for modern IF stages to become mistuned to the degree where a complete alignment procedure is necessary. If the need for alignment is clearly indicated, it can be performed after tuner tracking has been checked. Since the IF alignment, however, is a function independent from tuner tracking, the IF can be aligned without tuner tracking. It must be remembered that in television systems the initial sound carrier of 41.25 MHz is heterodyned with the picture carrier of 45.75 MHz to produce a new sound IF frequency of 4.5 MHz. Thus, IF alignment involves not only the IF stages that follow the tuner but also the second sound IF stages that follow the video detector and are applied to the FM detector.

In IF alignment, as with tuner tracking, the list of basic precautions given in Sec. 9-3 should be carefully studied before attempting IF alignment. The general placement of the sweep generator and scope for video IF alignment is shown in Fig. 9-3. Here the sweep generator is coupled to the input of the mixer stages. It is assumed that the sweep generator has a built-in provision for producing a marker pip. The vertical input to the oscilloscope is obtained from the video detector output load resistor, as shown. The shunting capacitor is sometimes necessary to bypass transient noises that may interfere in the obtaining of a clear visual image on the scope screen.

The sweep generator signal is adjusted to scan over the entire IF bandpass, which was illustrated earlier in Fig. 1-10(b). The marker pip adjustment knob is then turned to move the marker pip along the response curve. If the sound and video signal frequencies are not as indicated in Fig. 1-10(b), the alignment tuning devices must be adjusted to reposition the IF response bandpass to meet the frequency conditions required. Alignment of the 4.5-MHz sound IF is performed in similar fashion to that just described for the video IF alignment. As shown in Fig. 9-4, the input signal from the sweep generator is applied to the first sound IF circuit, and the scope is attached to the output of the FM

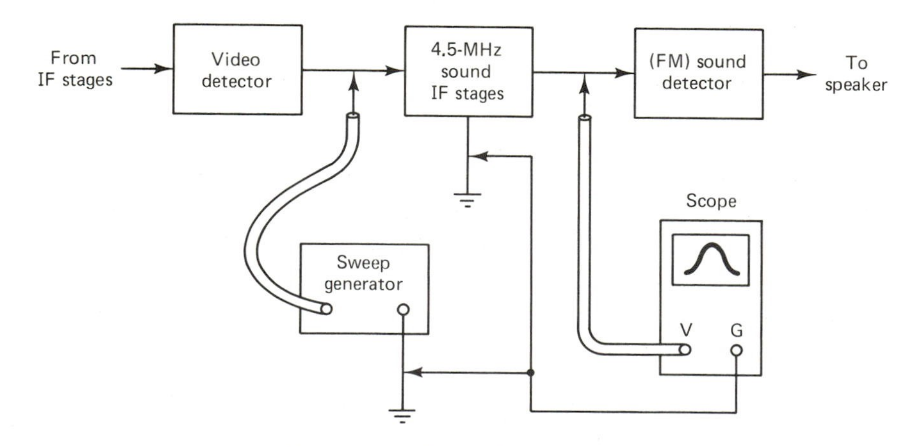

Figure 9-4 Sound IF alignment.

video detector. Now, however, a much narrower bandpass prevails since the sound carrier of 4.5 MHz has a maximum deviation of only 25 KHz on each side of center frequency. Hence, the total bandpass need only be 50 kHz.

9-5. TRAP ADJUSTMENTS

In Chapter 1, Fig. 1-13(a), three resonant circuit traps were shown between the output from the tuner system and the input to the IF stages. The trap portion of that illustration is repeated in Fig. 9-5(a). Here the first trap is designated as 39.75 MHz, the second as 47.25 MHz, and the third as 41.25 MHz. Since each trap is tuned to a separate frequency, it is unimportant in what sequence they are placed in the circuit. Thus,

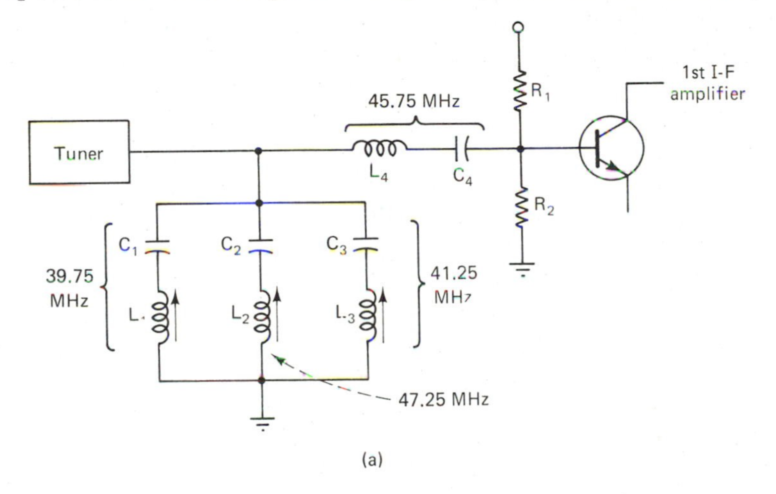

(a)

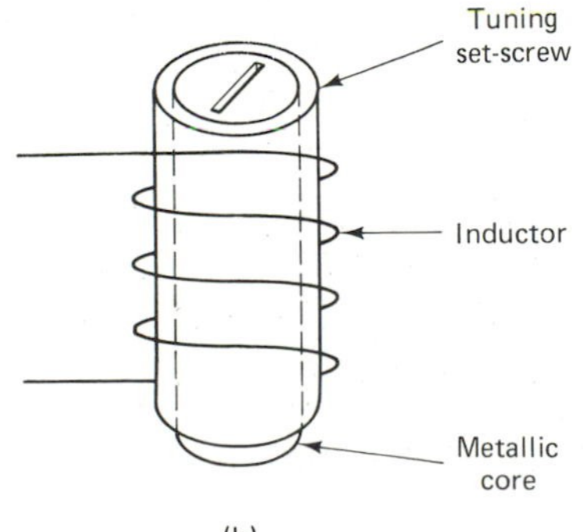

(b)

Figure 9-5 Interference traps.

another manufacturer may have designated the first trap as 47.25, the second as 39.75 MHz, and so on. These traps function on the principle of combining a capacitor and inductor in series to form what is termed a *series resonant circuit*. The latter has the characteristics of having a low impedance for any signal to which the circuit is tuned. Thus, the fist trap that is tuned to 39.75 MHz will have a shunting effect for this frequency and thus will minimize any interference which might be caused by an adjacent channel with a signal strong enough to be picked up. Similarly, the second trap is also tuned to an adjacent-channel intermediate frequency and hence will have a shunting effect and diminish it. The last series trap shunts some of the 41.25-MHz signal to ground and thus diminishes it sufficiently to prevent interference.

The adjustment of these traps is important to reduce adjacent-channel interference. Adjacent-channel interference can vary, depending on atmospheric conditions, even though adjacent channels are usually allocated to distant areas from a local station. Proper trap tuning is essential even in an area free of adjacent-channel interference. If the traps are misadjusted, they can interfere with the IF bandpass. Consider, for instance, the IF bandpass shown in Fig. 9-6. Here the adjacent-channel sound trap of 47.25 MHz is mistuned and consequently places a curvature dip near the IF frequency of 45.75 MHz. Consequently, poor picture quality will result. Compare this bandpass waveform with the desired waveform shown earlier in Fig. 9-2. With an improperly adjusted sound track interference bars will also appear on the screen. These bars will appear intermittently as the sounds in the adjacent channel vary. Interference from an adjacent-channel picture will cause faint images to appear on the screen. The interference images will be visible to the degree of interference. For a station farther away the images may be very faint. An interfering video signal may often float horizontally or vertically because it will not be perfectly synchronized with the vertical and horizontal sweep frequencies.

In Fig. 9-5 a special function is formed by the series inductor L_4

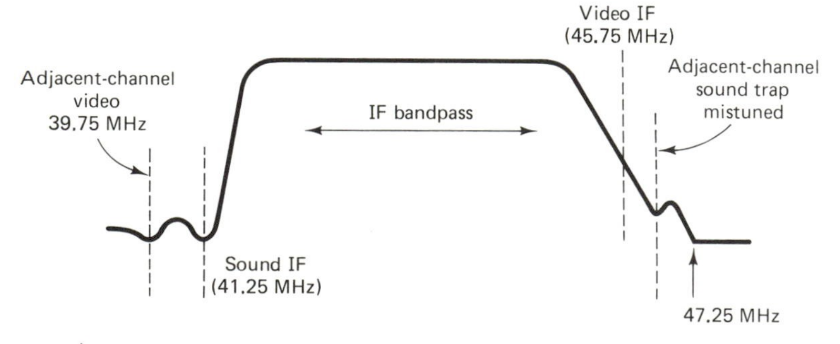

Figure 9-6 Mistuned sound trap.

and the series capacitor C_4. These two components also form a series resonant circuit and hence offer a low impedance for any signal to which it is tuned. This resonant circuit, tuned to 45.75 MHz, serves as a link between the signals from the tuner and the first IF stage. Thus, its low impedance assures passage of the picture IF signal but passes a lower-amplitude portion of the sound IF signal. Other signals having frequencies above and below the resonant frequency of the inductor combination find a high impedance and thus are attenuated.

Note that the three trap circuits shown in Fig. 9-5 show an arrow besides each inductor. The arrow represents a variable core which can be adjusted for tuning purposes. The adjustable core is in the form of a molded ferrite material, with the top usually formed with a slot for screwdriver adjustment, as shown in Fig. 9-5(b). In adjusting the tuning of any such traps or resonant circuits, a nonmetallic screwdriver should be used. A metal screwdriver can affect magnetic fields and alter the tuning. Thus, the circuit may be tuned perfectly while the screwdriver is in place, but as soon as it is removed, the core characteristics are altered.

Generally, such circuits, once tuned, hold their characteristics for a long time. If readjustments are necessary, they are usually only of a minor nature, requiring very slight touch-up tuning. If the traps are mistuned considerably because of having been adjusted carelessly or replaced with new components, a sweep generator and oscilloscope will be necessary. The latter should be set up in similar fashion to that for IF alignment purposes, as discussed earlier in Sec. 9-4 and illustrated in Fig. 9-3. Each trap should then be adjusted to obtain proper IF band-pass as identified by the marker pip, as illustrated in Fig. 9-2. The band-pass resonant circuit composed of L_4 and C_4 is preset at the factory and will only be mistuned if the inductor or capacitor become defective. The capacitor C_4 may open in which case the picture is lost. If C_4 shorts, it alters the tuning, and picture contrast declines. The inductor L_4 rarely causes problems, though it should be checked for any shorted turns or open-circuit conditions.

Another trap which needs to be adjusted on occasion is the 4.5-MHz trap located in the video amplifier section following the video detector. As shown in Fig. 9-7(a), the 4.5-MHz sound IF developed in the detector is sent to separate IF amplifiers. As pointed out earlier, the 4.5-MHz frequency is obtained by mixing the 41.25-MHz sound IF signal with the 45.75-MHz video signal. The difference frequency thus obtained is 4.5 MHz, which comprises the new and final sound IF signal. Once this is channeled to the respective IF stages and FM detector, its presence must be eliminated from the video amplifiers to prevent interference bars, which may be caused by such a signal. Hence, a series resonant trap is again utilized consisting of C_2 and L_4, as shown in Fig.

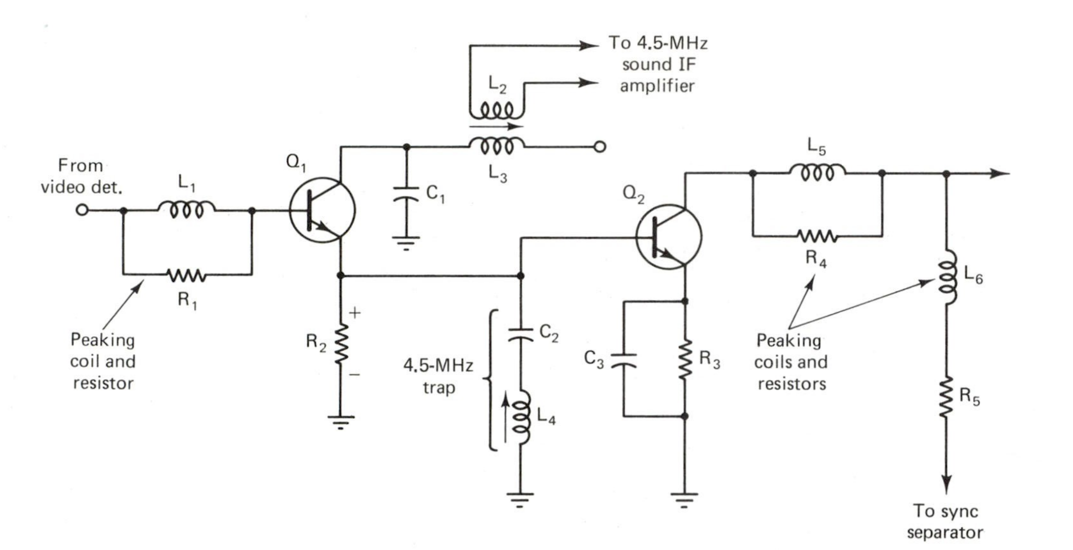

(a)

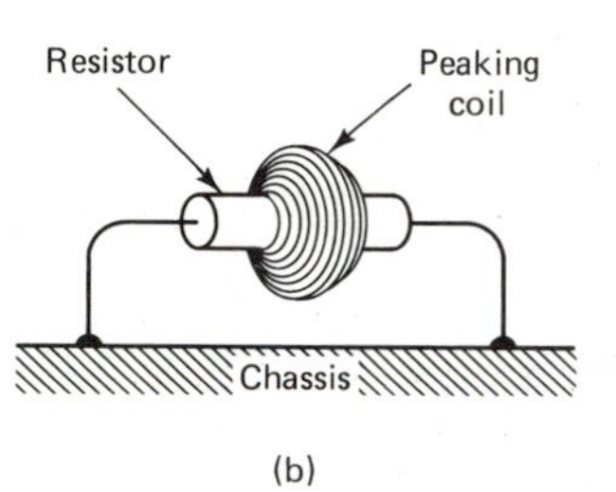

(b)

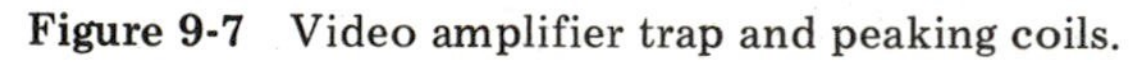

Figure 9-7 Video amplifier trap and peaking coils.

9-7(a). This trap can be adjusted without special equipment since its sole purpose is to minimize sound interference from the channel to which the receiver is tuned. Thus, if sound bars appear and they are intermittent and appear to be synchronized with the voice or music in the picture, the trap can be adjusted until the interference disappears.

Sound bars that appear on the screen sometimes identify the source by the nature of the bars. A heavy single horizontal black bar indicates 60-Hz interference and usually is caused by an open or defective filter capacitor that permits hum signals to ride through the circuitry. As the frequency of an interfering signal increases, the number of horizontal hum bars will increase proportionately up to the horizontal sweep frequency of 15,734 Hz, the horizontal sweep frequency for color receivers. Interfering signals above this frequency will cause vertical bars to appear on the screen. As the interfering frequency rises, the number of bars will increase.

9-6. LOSS OF PICTURE DETAIL

The video amplifiers shown in Fig. 9-7 are representative of the essential components necessary for processing picture and sound information. In their basic concept such amplifiers are essentially audio amplifiers, though their signal-frequency capabilities have been expanded. Where audio amplifiers need only have a response from approximately 30 Hz to 20 kHz, the video amplifiers must handle a frequency range extending above 4 MHz. The necessary wide-band response is obtained by using special inductors known as *peaking* coils. The latter are connected in series with the signals, and for Fig. 9-7 the peaking coils are designated as L_1 and L_2. Such inductors, in conjunction with the series capacitance between the elements of the transistors, form a high-pass filter network that extends the bandpass into the upper-signal-frequency regions necessary for sharp detail in video signals. Because a circuit composed of an inductor and capacitor in series forms a resonant circuit with a low impedance, it will have an increasingly higher impedance for lower-frequency signals. Thus, in essence, the frequency response of high-frequency signals has been improved. The resistor associated with the peaking coil broadens the frequency response and thus increases the effective range of such coils.

Peaking coils are often wound around the resistors that shunt them. Thus, the assembly shown in Fig. 9-7(b) is the general appearance of the combination of inductor L_1 and resistor R_1 as well as L_5 and R_4. The peaking coil and its shunting resistor combination are usually mounted so the combination is slightly above the chassis. By using longer leads than normally employed for other components, the peaking coil and resistor are spaced sufficiently away from neighboring metal parts, shields, or other similar components. The additional spacing minimizes the loss of high-frequency signals that could occur when the peaking coil and resistor form a capacitance characteristic with nearby metal objects. Any such stray capacitance has an increasingly lower reactance for successively higher frequency signals and hence will diminish them and counteract the benefits of the peaking coil and resistor. When replacing defective peaking coils, it is important to install the new ones with the same space between them and nearby objects as was the case originally.

Shunt peaking coils are also used, and a typical example is L_6 in Fig. 9-7. Here the inductance of L_6 shunts existing circuit capacitances (such as those existing between the inner elements of a transistor or formed by connecting leads to nearby metal objects). Thus, inductor L_6 forms a parallel resonant circuit with circuit capacitances and hence has a high impedance for signals having a frequency at or near the resonant frequency. The result is a high impedance that minimizes high-frequency signal losses that would occur because of the low reactance of circuit capacitances. The peaking coil L_6 has a series resistor R_5 that connects to the sync separator circuit. The collector voltage is also present in this circuit.

If series peaking coils L_1 and L_5 become defective or if the shunting resistor becomes an open circuit, the picture quality will deteriorate because of the reduction in high-frequency response. The latter causes a loss of fine detail in pictures because the high-frequency signals are the ones that produce the sharp detail. Since some peaking coils carry portions of the collector current, an open shunting resistor increases current through the peaking coil. Similarly, if the resistor opens, the full collector current will flow through a series peaking coil such as L_5 in Fig. 9-7. Sometimes an open circuit will occur at the place where the peaking coil terminals are soldered to the resistor terminals. When any doubt exists, one side of the peaking coil can be disconnected from the resistor and the circuitry. With an open end for both L_5 and R_4, for instance, a continuity check can be made individually for the inductor and resistor. This procedure discloses the shorted or open-circuit condition and is preferable to making checks while the components are connected into the other circuitry. There are some checking devices that will permit in-circuit component testing, but this procedure must be carefully done and is not recommended for all checks.

If the picture does not appear to have fine detail, other stations should be checked to make sure the trouble is not with the station being received. If the VCR is used for playback of new or rented tapes, other tapes should be tried. If the picture quality remains poor, the fault could lie with either the VCR or the television receiver. If possible, another television should be tried in this case. If a different receiver or monitor still has poor detail, it is likely that the fault lies in the VCR circuitry. A degraded picture that suffers from high-frequency signal loss could also be caused by a misadjusted focus control if one is present on the receiver. The tracking control on the VCR should also be checked for the proper setting, particularly if the tape being played is not recorded on that VCR. The loss of picture detail is a symptom that remains fixed in most instances and hence differs from the nature of interfering signals that intermittently cause streaks or lines of interference on the picture being viewed.

9-7. GENERAL VIDEO AMPLIFIER TESTS

There are many fault symptoms appearing on the screen that give a clue regarding the circuit or section that is causing such a defect. If a picture is present but no sound appears, it could be caused by faults in the circuits following the 4.5-MHz sound take-off transformer shown in Fig. 9-7 (L_2 and L_3). Note the arrow indicating tuning in the transformer for the sound IF take-off section. This is adjusted for maximum signal output. If this variable core is mistuned too much, it will cause a loss of sound. If the VCR and monitor have been reproducing sound satisfactorily and the sound output stops suddenly, it would not be the fault of the mistuned L_2-L_3 section. Hence, this control should not be adjusted in an effort to obtain sound if the sound suddenly stops. If this control is turned needlessly, it can impair sound output even after the true fault had been found and corrected.

If the sound output is all right but no picture appears, the fault may be in the circuits following the first video amplifier. Since the sound take-off functions sufficiently to produce audio, it is likely transistor Q_1 is functioning properly, unless resistor R_2 has shorted. If this resistor had opened, the emitter ground return would have been broken, and Q_1 would have become inoperative. *Note*: Lack of a picture in this instance assumes that the picture tube still lights up. If the picture tube is totally dark, it could indicate a loss of high voltage or a defective tube.

Loss of picture but presence of sound necessitates a careful reading of the voltages present on the collectors of each transistor. If the voltages are present at these places, there should also be a voltage drop across each emitter resistor (R_2 and R_3). Lack of voltage across R_2 would indicate a shorted resistor (which is unlikely with carbon-type resistors) or a shorted trap capacitor C_2. If there is no voltage reading across R_3, it could be caused by a shorted bypass capacitor C_3. If all components have been checked and found to be satisfactory, the transistors should be checked. If the circuitry shown in Fig. 9-7 checks out satisfactorily, other circuits must be tested as well as low and high voltages associated with the picture tube. (See also Chapter 8 and 10.)

9-8. COMB FILTER

A circuit section termed a *comb* filter is used to minimize cross-talk interference wherein undesired signals can cause faint though distracting double images. The filter is used in the chroma and burst section, as shown in Fig. 9-8(a). The filter principle is based on the concept of blending the interfering signals with identical signals 180° out of phase. The result is cancellation of the cross-talk interference. The process is similar to inverse feedback in audio systems to cancel undesired signals.

The filter samples the chroma signal and delays it by one horizontal line. The delayed signal is added to the undelayed signal in a resistive network, as shown in Fig. 9-8(b). For any evidence of cross-talk-type interference on the viewing screen the comb filter may be defective and must be replaced. Initially, however, tests and checks

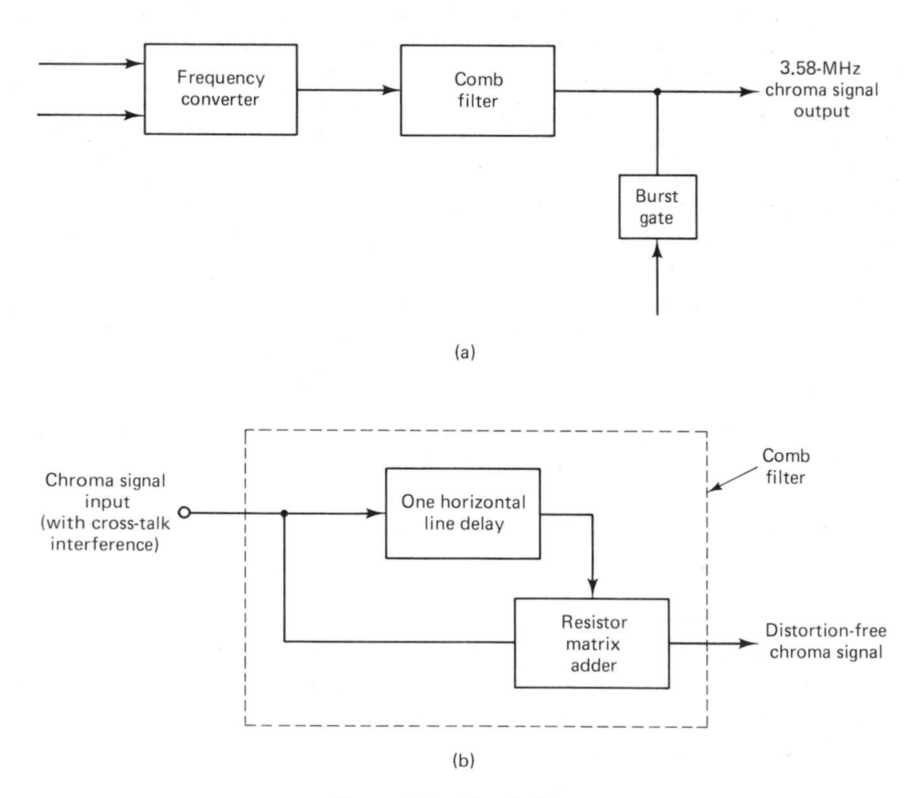

Figure 9-8 Comb filter.

must be made to make sure the problem is not caused by a tape recording that contains the interference. A badly worn tape, dirty recording heads, or magnetized heads may cause sufficient picture interference to mislead the technician into making an incorrect diagnosis. Both the chroma and burst signals must be of proper amplitude as checked on a scope and compared to the amplitudes specified on the schematic for the particular VCR being checked.

10

Defective Parts Isolation and Replacement

10-1. SIGNAL TRACING

In Chapter 2 the localization of defective linkage sections in VCRs and television sets were covered. Applications of test equipment were covered in Chapter 7, and the value of voltage and continuity measurements were stressed. In particular, continuity checks with the power shut off to the VCR helped isolate and pinpoint defective units or sections. Similarly, in Chapter 9 VCR sectional testing of tuning and alignment was discussed. Of great importance, however, is the procedure known as *signal tracing* for localizing the section that is defective. Once the section of the VCR has been isolated, additional checks (including continuity and voltage measurement) are then undertaken to pinpoint the exact component which may be causing the trouble. Thus, if the sections involved are printed-circuit boards with transistors, resistors, capacitors, and coils mounted on them, the defective unit can be replaced. If, however, a signal-tracing procedure pinpoints an integrated circuit as the cause of the trouble, an exact replacement would have to be obtained to restore operation.

In signal tracing we utilize an instrument that can sense the presence of a signal and follow this signal through an electronic circuit from its entry point to its exit. Thus, we can check for the presence of a signal in the antenna transmission line and localize the point where it enters the tuner section of the VCR. We then check for the presence of the signal at the output of the tuner (with appropriate tuning changes in the testing device being utilized). If the signal is present, it can then be followed through the intermediate-frequency stages and to the detector output. If the signal is no longer in evidence at any place along this route, the defective section is then identified.

In the preceding example the normal incoming signal was utilized as the signal to be traced through the circuitry. Another method is to inject a signal from a signal generator into the circuitry and then follow this signal through the successive stages. Thus, the signal generator can be left at the tuner input and the signal's presence checked at various points as it progresses through the VCR circuitry. An alternate method is to inject the signal from the generator into the last stage initially with the signal sensing instrument remaining fixed at the output. The signal generator is then progressively moved to the input, and where the signal disappears is then identified as the faulty circuit. The third alternative is to inject a signal into each stage and sense its presence at the opposite of that particular stage. The advantage of any one of these methods over the others depends on the nature of the fault, the presence or absence of an incoming signal, and other factors as discussed in succeeding sections.

10-2. AUDIO SECTION TESTING

An audio amplifier section presents few difficulties in troubleshooting and in the isolation of defective circuits and components. Since the signals handled are not in the RF range, the precautionary problems associated with diagnosing RF circuit faults are minimized. The audio amplifier stages that follow a detector usually consist of a preamplifier circuit, followed by a conventional transistor amplifier and finally the power output system. For purposes of illustration, assume that we have a defective audio system of the type illustrated in Fig. 10-1. Here a volume control to the base terminal of a transistor input is illustrated. The representation could also have been a triangle indicating a silicon chip. Here the output is applied to a speaker as in a television receiver or monitor.

If this amplifier system feeds the output terminal of a VCR, there would be very little power amplification and hence not enough to drive a speaker. For an inoperative audio system it is essential that we isolate the offending stage so that component checks are held at a minimum. As mentioned in Sec. 10-1, several signal-tracing procedures can be utilized. If we are testing a circuit with a loudspeaker, the latter can then be used to check for an output during signal tracing. In one procedure we could apply a signal generator to the input of the transistor circuit, as shown. The volume control must be placed at a maximum setting so the input circuit is not grounded out. The signal generator

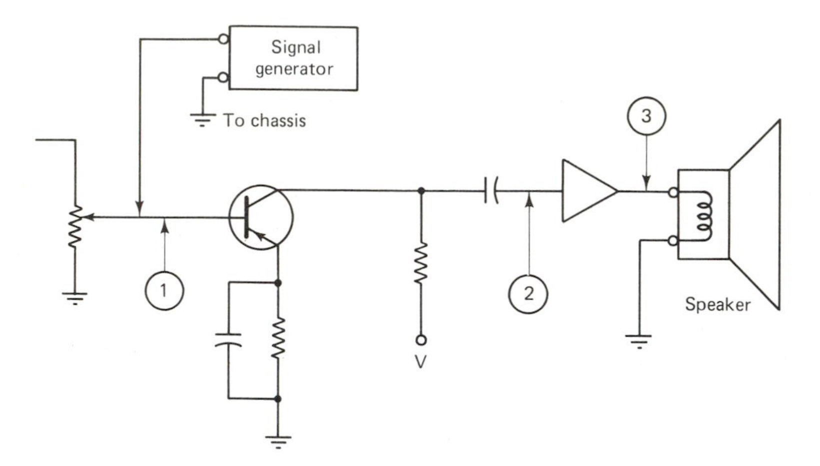

Figure 10-1 Signal injection.

should be set at an audible frequency such as 400 Hz. If this sound is now heard from the speaker, it indicates a signal path through the circuitry. The output from the signal generator can be varied to make sure that the circuits handle the increased volume levels satisfactorily. Thus, if the audio stages work satisfactorily for this test but not during normal playback, it would indicate that the trouble lies in a circuit prior to the audio section. Hence, the IF stages or the tuner must be checked in monitors. For the VCR the output from playback tape heads must be verified.

If no signal appears at the output for a signal injection at test point 1 in Fig. 10-1, the signal generator should then be placed at test point 2. If a signal now appears at the output, it indicates the fault lies somewhere between test point 1 and test point 2. In such an instance the transistor and associated components will have to be checked, as detailed later in this chapter. Voltage readings must also be taken to make sure voltage distribution to all circuits has not been interrupted by faulty parts.

If no output is obtained when the signal generator is placed at test point 2, it indicates that the silicon chip is defective or that the speaker had an open coil. To check the latter, the signal generator can be placed directly across the speaker terminals. If a sound is now heard, the defect has been localized to the integrated-circuit (IC) chip. In such an instance an exact duplicate replacement must be obtained from the distributor for the particular system being tested. If no sound is forthcoming and the signal generator was placed across the speaker terminals, it could indicate an open voice coil or a broken connection in the line feeding the speaker. Check for poorly soldered interconnections between the output of the amplifier and the speaker. An extra test can be made by placing a separate speaker across the output temporarily to check for sound output.

With the foregoing procedures the signal generator is moved progressively from left to right, thus checking for signal progression through the system. As an alternative the signal generator could have been placed across test point 3 and then progressively moved to the left in reverse order from the foregoing.

10-3. TRACING IF AND DETECTOR STAGES

In Sec. 10-2 it was mentioned that if a signal input to the first audio stage produced an output, the fault would be in circuitry preceding that stage. In testing stages prior to the audio amplification, it will be necessary to use a signal generator capable of supplying a radio-frequency output signal. Such a signal can be injected into test point 3 of Fig. 10-2. This signal, however, would have to be frequency-modulated since this is the system utilized in television transmission and in the VCR or monitor. Most FM detector systems utilize the ratio detector circuit wherein two diodes, wired in reverse to each other, sense the frequency deviations of the modulated signal and produce an audio output. The signal feeding test point 4 is the resultant audio signal and can be displayed by an oscilloscope or can be heard from a speaker if one is present. (The audio output from a VCR can be heard with earphones.)

Failure of a signal output from the ratio detector calls for a check of the diodes, resistors, capacitors, and coils for finding an open- or short-circuit condition. If there is a sound output for a signal injected at test point 3, the signal generator can be placed at test point 2. Obviously, if no signal output is obtained, it would indicate a defective IF stage. Similarly, the signal generator can be applied to test point 1,

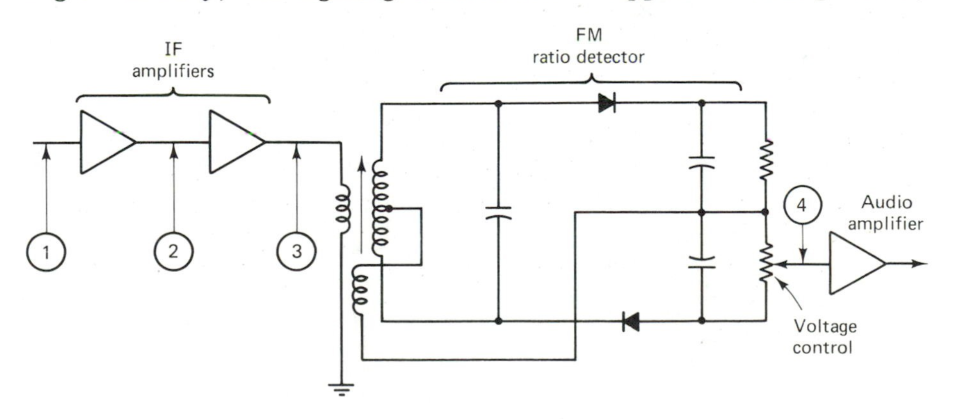

Figure 10-2 Signal tracing FM sound stages.

which is the input to the IF stages as well as the output from the tuner. An oscilloscope can remain in place at test point 4 or at the output of the audio amplifier as desired. With appropriate input plugs, the audio output from the VCR can be monitored by a scope or by earphones, as previously mentioned. If the IF stages are in chip form and an inoperative system is indicated, an exact replacement chip will have to be obtained to replace the defective one.

10-4. IC CHIP REPLACEMENT

There are two methods used in mounting a chip on a printed-circuit board. The most convenient (but the more expensive for the manufacturer) is the use of a socket into which the chip is plugged. Thus, a defective chip is simply unplugged and a new one inserted into the socket, as shown in Fig. 10-3(a). Care must be taken to insert the new chip with the identifying dot or indentation in the proper direction. Another method of chip mounting is to insert the chip into a series of prepunched holes on the printed circuit board, as shown in Fig. 10-2(b). In such an instance the prongs which extend beneath the chassis are

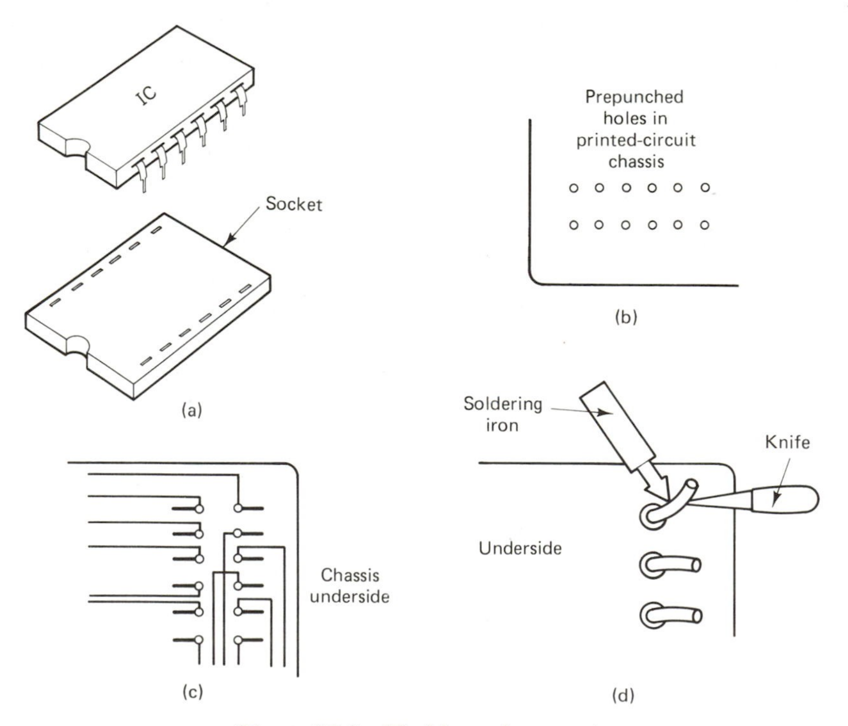

Figure 10-3 IC chip replacement.

bent over as shown in Fig. 10-3(c) and soldered into place. Obviously this process makes the removal and replacement of a chip very difficult.

To replace a soldered IC chip, extreme care must be taken during the unsoldering of the prong to prevent heat damage to the printed-circuit interconnections. Too hot a soldering iron (or maintaining the iron tip at a point too long) could melt the printed-circuit interconnections as well as the plastic foundation material of the printed-circuit chassis. Initially each prong on the underside must be unsoldered and lifted up, as shown in Fig. 10-3(d). It is preferable to use a small low-wattage soldering iron rather than the soldering gun type of unit which applies high heat very rapidly.

The soldering iron tip must be held on the prong for a sufficient length of time so that the prong can be lifted by a knife point, as shown. Some repair technicians use a small compressed-air blower or a hand-held rubber blower to rid the unsoldered section of loose solder that may prevent the chip removal. Each prong of the chip must be treated in a similar manner until all prongs have been unsoldered. The chip is then pulled out, though on occasion the soldering iron may have to be used temporarily at one or two prongs to loosen them additionally. Once the chip has been removed, make sure the replacement is inserted in the proper direction with respect to the identifying identation or dot at one end.

10-5. VIDEO AMPLIFIER CHECKS

The procedures for checking video amplifier stages are similar to those used for the audio sections discussed earlier in Sec. 10-2. For the video system an amplitude-modulation detector is utilized instead of an FM detector as for the sound, as shown in Fig. 10-4. The output of the AM detector is applied to the required number of video stages to bring the

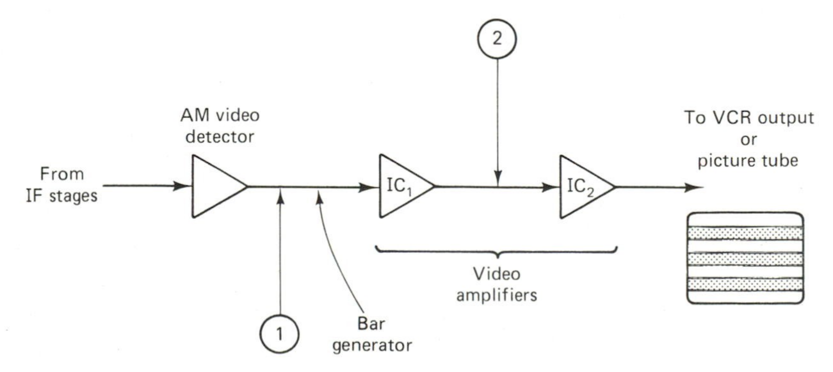

Figure 10-4 Checking video amplifiers.

signal up to the level required for application to the picture tube or to the video output terminal of the VCR. (The latter is not the cable output which feeds the antenna system of the television receiver but the auxiliary video output illustrated earlier in Fig. 2-6.) An audio signal generator or a bar generator is used as the input signal. (See Fig. 7-20.) With the bar generator input set to produce horizontal lines, the pattern should appear on the picture tube. If a signal appears when the bar generator is applied at test point 1, it shows the video amplifiers are functioning properly. If no signal appears at the output, the bar generator can be applied to test point 2. If a signal now appears at the output, it indicates that chip IC_1 is defective. In making such tests, the gain control should be at a normal setting and the output regulated by the bar generator. If the video amplifiers function normally during such testing but no picture appears during operation, the video detector and IF stages must be checked, as was the case for the audio amplifier circuitry. For the video detector test, use an RF signal set at the IF frequency and modulated with a 400-Hz tone. When the latter is applied to the detector input, it should produce bars on the screen if the detector is functioning properly. If not, the diode and other component parts of the detector must be checked individually to find the fault. As with the audio amplifier test discussed earlier, it is assumed that the unit operated satisfactorily and that the loss of video occurred suddenly. With weak signals rather than absence of signals, the fault may lie with mistuned stages, as discussed later.

Since the video information is amplitude-modulated, it resembles audio except for the much greater signal-frequency range. Hence, however, advantage can be taken of the detector's response to audio by sensing the output from the video detector with either a scope or earphones, as shown in Fig. 10-5. If the detector is functioning properly, the RF signal generator set at the IF and modulated with a 400-Hz tone can be

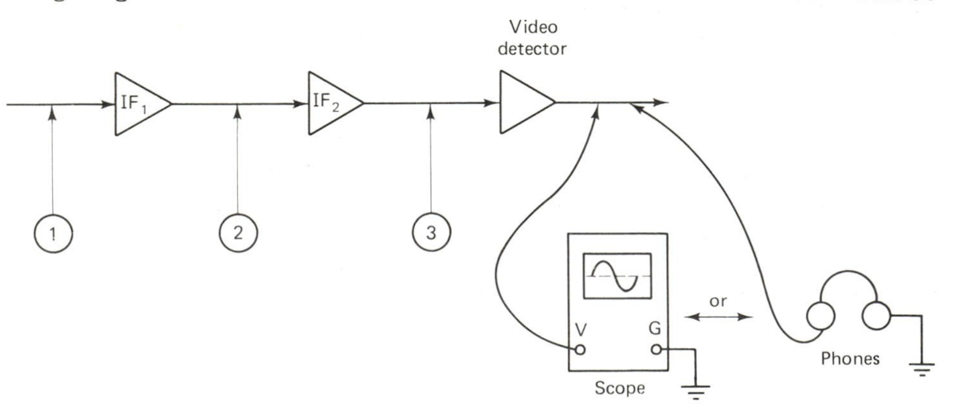

Figure 10-5 Video detector output sensing.

injected at test point 2 and the signal observed on a scope. If no signal is observed for a signal applied at test point 2 but a signal output is obtained for a generator input at test point 3, it indicates a defective IF_2 chip. If a signal is obtained at the output of the detector, the signal generator input can be applied to the input of the first IF stage at test point 1 for testing that stage. Since common IF stages are used for both the video and sound, a check of the IF stages thus tests for both the video and audio IF signals. It is only at the individual detectors that the resultant video and audio are separated into individual paths. (Reference should be made to Fig. 1-13, 1-14, and 1-15.)

10-6. TUNER TESTING

The general schematic representations for tuners were given earlier in Fig. 1-12. With the older tuners a mechanical wafer switch tier was utilized for station selection. With modern electronic tuners, however, varactor diodes are utilized for push-button or touch-tuning station selection. (See also Fig. 1-11.) If signal tracing of IF amplifiers, detectors, and output amplifiers have failed to disclose the circuit that is at fault, the problems may be caused by faulty tuners. In electronic tuners individual shielded containers may be used for the VHF and UHF sections, though the two can be contained in a single shielded compartment. Many tuner sections have test points available for signal injection or signal output checks as well as test terminals for voltage readings. Since the tuner is usually an individual section well shielded, input and output terminals are generally of the plug-in type, as shown in Fig. 10-6. With the signal-tracing procedures of the IF, detector, and amplifier stages completed, the lack of signal would indicate tuner problems.

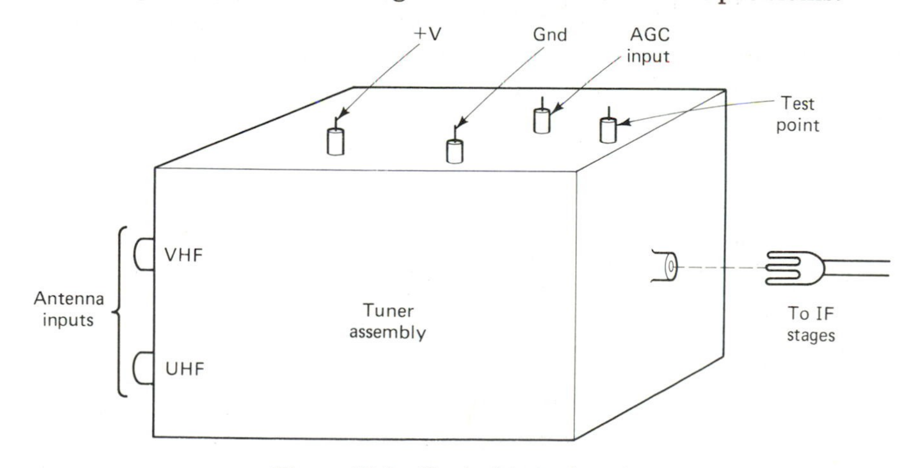

Figure 10-6 Typical tuner housing.

Before tuner signal tracing, all connections to the tuner should be carefully checked to make sure they are firmly in place or well soldered as the case may be. With voltages and connections verified as satisfactory, the input and output to the tuner must be checked. If the tuner is defective, both sound and picture would be lost. Unless a serious short circuit exists within the tuner that affects both VHF and UHF input, an inoperative VHF section may not necessarily alter good reception from a UHF input or vice versa. Since the antenna or input cable system furnishes a signal, the need for a signal generator is not necessary. The functional aspects of the antenna system can be checked by applying the antenna lead-in directly to a television receiver to be in good working order for verification. If good signals are obtained for VHF and UHF but no output is obtained when the lead-in is applied to the defective unit, it is possible that a defective component is present in the tuner. Again it is assumed that the trouble developed suddenly and is not the fault of improper tuning made during the station-selection or the fine-tuning set-screw adjustments for establishing the channel selection of the push buttons for touch-tuning devices.

If the trouble has been localized to the tuner, it may be necessary to replace the entire unit since tuner repairs are difficult and parts positioning during replacement may also be critical, particularly in the UHF tuner section. Also, since the tuner compartments are crowded with components, it is difficult to check and replace individual sections because it is a time-consuming process. As shown earlier in Fig. 1-12, resonant sections rather than the conventional inductors are used for tuning purposes in the UHF tuner. Failure of the transistor in the UHF tuner or any of those shown in the VHF section would cause complete failure and require replacement. The AGC (automatic gain control) input line could also affect the operation of the tuner if any of the resistors have opened. A thorough check of components includes a comparison of the resistance values obtained by an ohmmeter and those values given in the schematic for the particular model under analysis. Failure of the resistors or radio-frequency coils in series with input lines will also cause tuner failure, and these must be checked along with coupling capacitors. The latter must be checked for open and shorted conditions rather than for a change of value since the latter condition is rare.

10-7. AGC TESTS

Monitors and receivers of all types employ circuits that automatically control the gain of the video signal. In VCR units a form of such control is also used in the tuner and IF circuitry. The system is designated as automatic gain control (AGC), and a similar circuit is used in radios but is called automatic volume control (AVC). The HGC system regulates the gain of the video signal so that it remains substantially at the point where it had been set by the operator. Video gain represents a gain in contrast. Thus, if the contrast is increased, it automatically increases the gain of the signals arriving at the picture tube.

The AGC system samples a portion of the video output and applies this to an AGC transistor, as shown in Fig. 10-7. If the incoming signal from the antenna increases, the change is felt at the AGC circuit, and a correction voltage is applied to the tuner and IF stages. This voltage causes an automatic reduction in gain for those stages. Thus, the undesired increase in the incoming signal has been nullified by the AGC system. Similarly, if the incoming signal fades slightly, there is a decrease in video output, and the AGC automatically alters the voltage to the base terminal of the tuner and IF amplifiers to increase the gain of the video signal and compensate for the weaker incoming signal.

When defects occur in the AGC system, there may be sharp changes in the contrast level of the picture being viewed. A defective AGC system is much more evident when atmospheric conditions are such that the incoming signal fluctuates and has a tendency to fade in and out. To check the AGC circuit, tune the receiver to a local channel and measure the AGC voltage present at the input of the first or second video IF amplifier. A good-quality voltmeter should be used having high sensitivity and a minimal loading effect on the circuit being checked. When the receiver is tuned to other channels, the voltage being measured (termed the *bias voltage*) should change noticeably when different channels are tuned in.

If the bias voltage remains constant, the circuit must be checked for faulty transistors or components. Initially check the settings of the

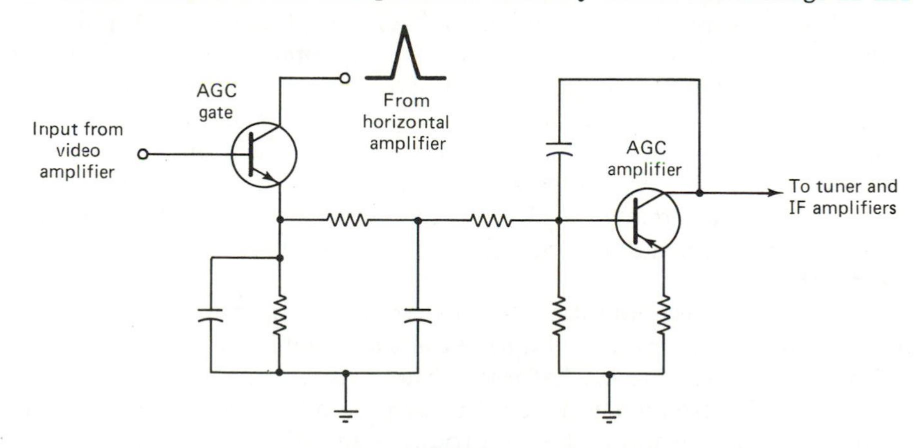

Figure 10-7 Basic AGC circuitry.

AGC control and note if it has any effect on contrast. Absence of bias can cause bending and tearing of the signal on the screen, indicating overloading. The presence of a strong bias signal that does not fluctuate will result in a faded picture in which the contrast cannot be improved.

The type of AGC in modern receivers is termed *keyed* AGC. In this system the AGC function is slightly delayed to minimize contrast changes due to transient noise pulses. The composite video signal from the video amplifier in conjunction with blanking is applied to the AGC circuit. The latter is so designed that only the sync tips have sufficient amplitude to initiate the AGC function. Thus, since the AGC operates only on sync tip levels, neither the video signal nor noise signals riding on them can influence the AGC function. Additional tests that may be required are voltage input checks as well as a test for changes in the resistance values of the various resistors utilized.

10-8. AFT SYSTEM CHECKS

Most modern monitors, receivers, and VCR units employ automatic fine tuning (AFT). The latter electronically adjusts the frequency of the tuner for proper resonance with the channel that has been selected. Thus, the system automatically adjusts the fine tuning without the necessity for constant turning of the knob type of tuning control. In many receivers and monitors the fine tuning is preset and automatically controlled so that no fine-tuning adjustment is necessary when changing channels.

The AFT system can be contained in a single chip, though on occasion a wired circuit may be encountered. Variations are found among the manufacturers, though basically the operational aspects are the same. The representative AFT system is shown in Fig. 10-8(a). The two diodes form the fine-tuning lock circuit that uses a 45.75-MHz input signal. When this phase discriminator obtains a signal of proper frequency, no voltage output is obtained. When, however, the tuner is mistuned and does not produce the proper 45.75-MHz signal, a correction voltage is produced at the output of the diode discriminator circuit because of the unbalance created. Such a correction voltage has a polarity and amplitude related to the nature and the degree of the mistuning (whether it is tuned above or below the proper resonant point and to what extent).

The correction potential obtained from the discriminator output is applied to a varactor diode shunting the tuner oscillator. The correction voltage thus alters the oscillator tuner and brings it into resonance with the 45.75-MHz frequency. When the correction is achieved, the discriminator is in balance and no longer produces an output voltage.

When station detuning occurs because of failure of the AFT system,

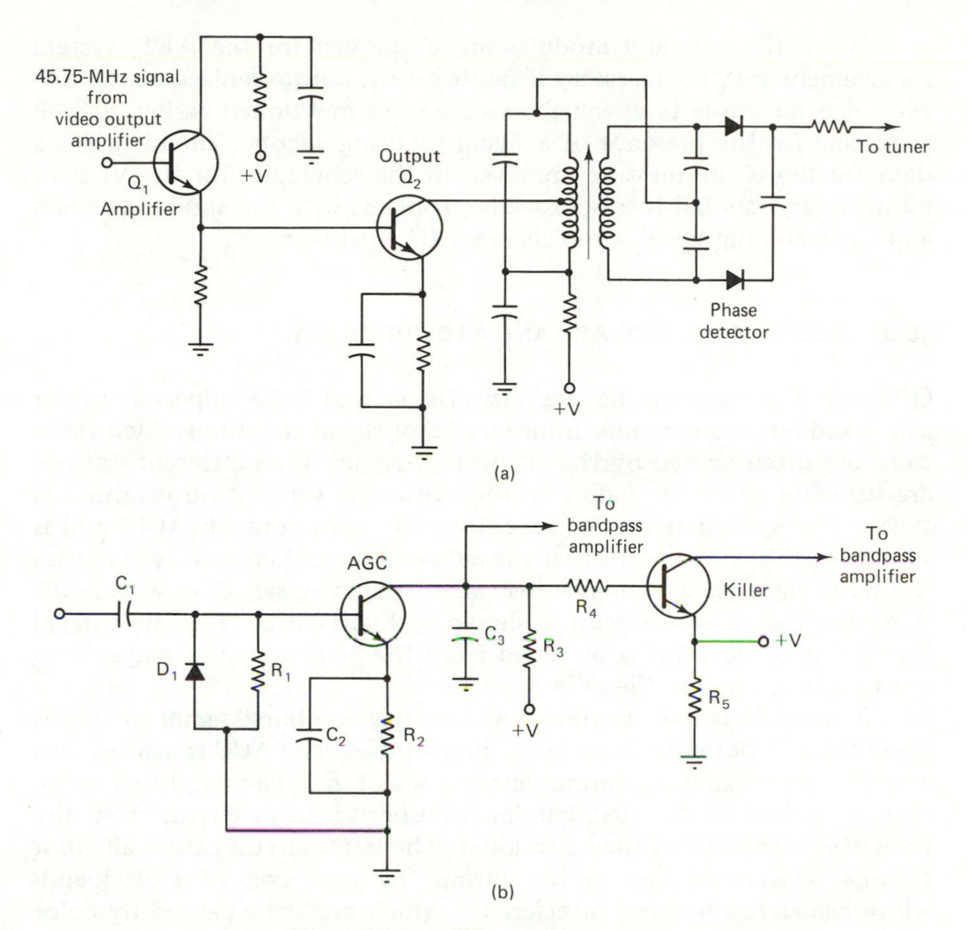

Figure 10-8 AFT and ACC systems.

voltage checks should be made to make sure proper potentials are applied to the collector elements of the transistors or to the representative voltage prongs of the integrated-circuit chip. If the voltages are correct, the transistors and diodes should be checked for short or open circuits. If these units are not defective, the components (resistors and capacitors) must be checked. The system should also be checked for the presence of a signal from the output of the video IF stages. In many VCR units preset channel tuning is performed by using a small tuning tool to adjust channel selector set screws. A switch is automatically disengaged when the control section lid is opened so that channel selection can be made without the AFT system in operation. Once the proper center-channel tuning has been made, the AFT circuit switch is automatically engaged to stabilize the reception of the particular channel to which the VCR is tuned.

When IC units and modules are employed for the AFT, system replacement may be necessary if faults occur. Before replacement, however, it is advisable to check the voltages, as mentioned earlier, as well as testing for the presence of a signal by using a scope. The identifying data for the IC in question are given in the schematic for the VCR or monitor. If tests fail to disclose any problems with the applied voltages and the incoming signal, a replacement IC must be tried.

10-9. TROUBLESHOOTING ACC AND ATC CIRCUITRY

Often used is a system that stabilizes the gain of the bandpass amplifier at a fixed level to minimize improper color signal variations. Such variations are often caused by the changes of signal when different stations are tuned in or by the fading of the station power as it travels through space. The system is known as automatic color control (ACC) and is similar to the AGC except that the automatic color control system uses the burst signal as a reference instead of the sync signals as with AGC. A representative ACC circuit is shown in Fig. 10-8(b). The input signal for the ACC transistor is obtained from the phase detector and subcarrier oscillator section. (See Fig. 1-16.)

The diode is used to detect and rectify the burst signal so that it becomes a dc potential for regulation purposes. The ACC transistor also amplifies the signal appearing across resistor R_3. The amplified signal also is applied to the first bandpass amplifier input circuit from the transistor identified as the color killer. The latter circuit causes all color circuits to become inoperative during the reception of a black-and-white signal to minimize interference which might be caused by color circuitry. In the absence of a burst signal the ACC transistor conduction drops, and the voltage existing at the junction of resistors R_3 and R_4 rises toward the maximum positive value. This potential becomes greater than that present at the emitter, and conduction occurs. The resultant bias change in the collector stops conduction of the second bandpass amplifier. With the presence of an input burst signal the proper degree of bandpass amplifier color signal gain is maintained automatically.

The ACC circuit shown in Fig. 10-8 must not be confused with the automatic tint control (ATC) circuit illustrated in Fig. 10-9. The latter maintains a constant flesh-tone characteristic during color reproduction. It eliminates the need for constant adjustment of the tints or hue controls for changes in camera levels or the slight differences in tint that may occur for different channels. As with the ACC, the entire ATC system can be incorporated within a single integrated-circuit chip. Although variations in the design of ATC systems are numerous, the one shown in Fig. 10-9 serves as a representative one for understanding

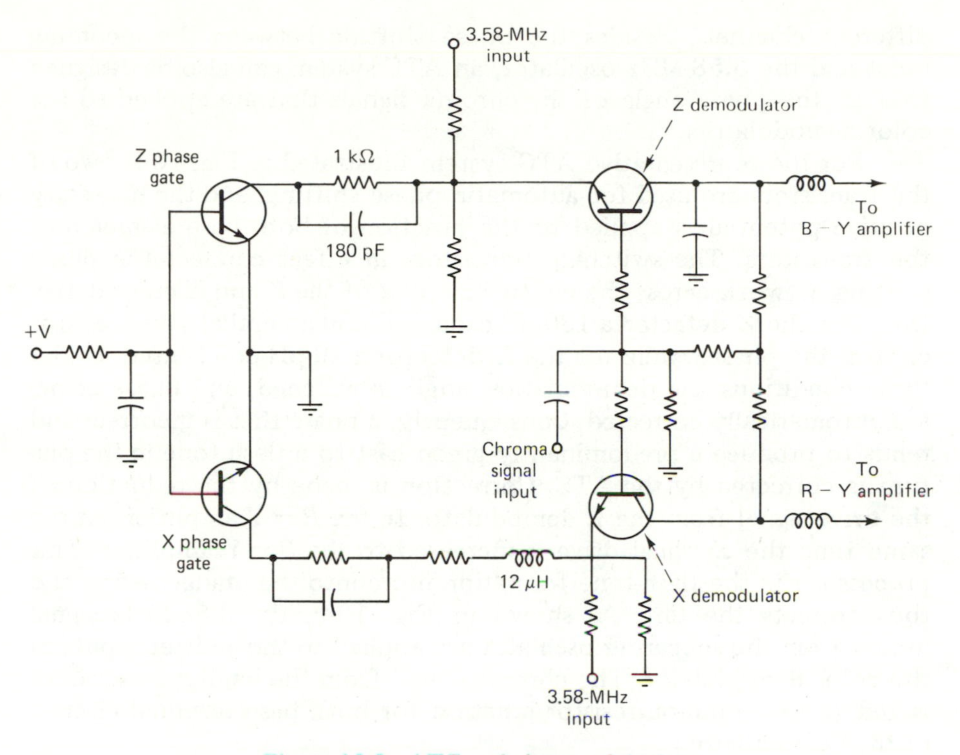

Figure 10-9 ATC and chroma detectors.

the basic functions and as a guide to the type of checks and tests that can be performed. (The Z and X demodulators shown in Fig. 10-9 are covered more fully in Sec. 10-15.)

The ATC circuitry is designed on the principle that every color that appears on the television screen is directly related to the difference in phase between the color signal and the 3.58-MHz burst. (The latter represents the exact frequency of the color signal subcarrier.) With the $R - Y$ signal at 90° and the $B - Y$ at 180°, natural appearing flesh tones in the TV picture are limited to a rather narrow segment of the color circle given in Fig. 7-24. Thus, such flesh tones range only a few degrees around the 57° point. When the tint control of the color television receiver is varied, it alters the phase relationship between the incoming burst signal and the 3.58-MHz oscillator signal of the phase-locked loop. (The latter was shown earlier in Fig. 1-16 and consists of the phase detector, the reactance control, and the 3.58-MHz oscillator.) The tint control thus makes it possible to adjust flesh tones between a range having a greenish color at one end and an excessive amount of red at the other. With the ATC system, however, there is no longer a need for resetting the tint control to correct for variations in transmission from different channels. Besides the phase shifting between the incoming burst and the 3.58-MHz oscillator, an ATC system can also be designed to shift the phase angle of the chroma signals that are applied to the color demodulators.

For the representative ATC system illustrated in Fig. 10-9, two of the transistors are used for automatic phase shifting, and the necessary positive potential is applied to the junction of both base elements of the transistors. The switching transistors in effect connect the phase shifting network across the emitter circuits of the Z and X color detectors. For the Z detector a 180-pF capacitor and a parallel 1-kΩ resistor control the phase, while for the X detector a 10-pH coil is used. Under these conditions the demodulation angle is widened, and phase errors are automatically corrected. Consequently, a phase that is incorrect and tends to produce a predominantly green cast to a flesh tone in the picture is corrected by the ATC. Correction is made by the reduction of the drive signal from the X demodulator to the $R - Y$ amplifier. At the same time the Z signal drive is increased to the $B - Y$ amplifier. This process shifts the flesh-tone formation more into the orange region and thus corrects the tint. As shown in Fig. 10-9, the 3.58-MHz signal inputs from the subcarrier oscillator are applied to the emitter inputs of the color demodulator. The chroma signal from the bandpass amplifier is fed to the common resistor junction for both base terminal circuits of the demodulator.

Once the tint has been adjusted by the manual control, the ATC circuit should function without the need for readjustment. If tint variations are noticeable, the improper flesh tones may be the result of a maladjustment of the camera and transmitter of the particular channel being viewed. Thus, other stations should be checked. If the antenna system feeds through the VCR, there should be no disturbance of tint since the video and audio carriers are transferred directly. If, however, station selection is performed by the VCR tuner, a misadjustment of the tuning for that particular channel may be at fault. Again other channels should be viewed before changing tint control settings.

If the flesh tones are incorrect for all channels, readjustments of the tint control should be undertaken. If proper flesh tones cannot be achieved, the trouble may be caused by faults in the ATC system. As with other circuits, the initial checks should consist of voltage readings to make sure proper potentials are supplied to the ATC circuits and transistors. A schematic for the particular monitor or receiver would be useful since it would specify the exact voltages that should appear at the terminals, and the schematic also specifies component values (resistors, capacitors, and coils). If components and voltages are correct, the transistor must be checked and replaced if found defective. For an IC, of course, an exact replacement must be obtained.

10-10. GRADUAL PICTURE DETERIORATION

In earlier sections problems associated with various control systems were discussed. For most of them it was assumed that the VCR, monitor, or receiver utilized in conjunction with a VCR operated normally and then suddenly developed problems. There are many occasions, however, where a slight mistuning occurs due to the gradual deterioration of some components (such as capacitors). Often humidity factors or temperature variations will affect peak performance, and in time sufficient picture degradation occurs to be noticeable. If no component has become defective or if the characteristics of the component have not changed too much, correction often consists simply of retuning or readjusting set screw or knob controls.

Once a VCR and monitor have been adjusted properly, few changes occur in color phasing or convergence. As transistors age or are replaced, slight touching up may be required. Video IF alignment, tuner tracking, convergence, and sound IF alignment are, in most instances, not required. In particular, tuner and IF alignment should not be undertaken unless the need for it has been definitely established. If the set screws that adjust alignment are haphazardly turned in an effort to regain sharpness or picture purity, it will be extremely difficult to reset the controls properly without using sweep generators and other equipment, as detailed in Chapter 9. Unless the proper equipment is utilized initially and the alignment procedures are properly followed, they should not be attempted because they may aggravate the existing faults.

Incorrect colors for all channels require resetting the color and tint controls. A check should also be made of the fine-tuning adjustment if this is available in manual control. Incorrect antenna orientation may also provide poor reception. An image that displays an undesired predominant color regardless of the station to which the unit is tuned requires a readjustment of the color gain controls. Proper convergence and proper levels of amplification for the three signals of red, blue, and green are essential requirements to achieve good color rendition and also to obtain pure white images on the screen when these are part of the scene. As a quick check the color control can be turned down completely and the resultant black-and-white picture analyzed. Colored areas should no longer be visible, nor should color fringing around scenes in the black-and-white picture. Such false colors indicate maladjusted color gain controls or incorrect convergence.

If pictures are not sharp, the fault may be an improperly adjusted focus control if one is available on the rear panel. If no focus control is present, a check will have to be made of the voltage applied to the focus electrode of the picture tube to make sure it is of correct value. If a focus control is present, precise adjustment is facilitated if you closely observe the horizontal trace lines of the screen. The lines should become sharper as the focus is improved. Temporarily, the vertical hold control can be misadjusted slightly to pair up the dual field lines and thus make them more visible on close inspection. Once good focus is obtained, the vertical control is again set for good stability.

A common trouble is the appearance of overlapping images on the screen. Double images may be caused by adjacent-channel interference or by ghost reception caused by the antenna picking up reflected signals from nearby buildings. Ghost reception can be minimized by reorienting the antenna to slightly different positions or by using an antenna with a sharper and more directional pickup pattern. Problems of this type occur more often during UHF reception, and greater care must be utilized in proper antenna installation and orientation.

Many antenna problems can be minimized by installing an antenna rotating unit that helps pinpoint the antenna for best reception. Low-loss lead-in lines should also be used to minimize interference. For the twin-lead type of transmission line, use insulated stand-off mounts and do not run such a line within a few inches of a metal area such as a rain pipe, telephone line, or electric line. Do not leave more than a minimum amount of slack in the transmission line behind a VCR or receiver. Leave only sufficient slack so that the VCR or receiver can be moved for cleaning purposes. When more than a few feet of transmission line are coiled in back of the VCR or TV, considerable loss in reception is introduced.

The cables that convey pay television to the home are of the coaxial type. Although these have a greater loss factor per given length than twin lead, they provide an outer shield for the inner signal conductor and therefore minimize interference pickup. Hence, any picture interference or deterioration that is obviously part of the signal sent via the cable system should be called to the attention of the cable company for correction. For coaxial cable lines, however, it is also preferable to have a minimum amount of slack behind the receiver to minimize cable losses.

10-11. PARTS REPLACEMENT FACTORS

When signal-tracing procedures have localized a defective circuit in troubleshooting procedures, the next step is to pinpoint the exact component which has become defective. Once the resistor, capacitor, or coil has been identified, it must be removed and replaced with an exact duplicate if at all possible. There are a number of factors, however, which must be considered in replacing components in a VCR or monitor. In replacing any type of coil, care must be taken to position the coil in the same manner as the original. Since coils tend to generate magnetic and

RF fields, they may interfere with other components. Some coils are shielded to minimize the problem, but incorrect positioning may still nullify some of these benefits. In some instances improper positioning (such as a power transformer) can introduce considerable hum or other problems. Often just a slight shift or rotation may cause some interference. This factor must be considered primarily when an exact replacement is not available and new mounting holes must be formed in the chassis.

Coils for tuning purposes in the RF regions or for filtering such as the radio-frequency choke coils should be exact replacements of the microhenry or millihenry values. Tuning coils are particularly critical, and even a slight value deviation can cause problems in reception and performance. Coils in some of the high-voltage circuitry are not only critical in value but also in the type of shielding and their replacement since they can increase X-ray radiation above the preset level established during manufacture. Schematics for monitors and television receivers usually indicate the sensitive areas where replacement by exact duplicate components is essential.

In some areas the value of resistors may not be too critical, and they may have a tolerance rating of 5 to 10%. Thus, a 100-Ω resistor may often be replaced by one having a value ranging from 90 to 110 Ω. Unless such a tolerance is specifically given, a replacement should be as close as possible to the original value. In some circuits where tuning or critical voltage values are involved, the tolerance is often expressed as 1%. Another consideration is the wattage rating of resistors. In power supplies or in power supply lines feeding amplifiers, some resistors may have a rating of 1W or more, and in some instances a wire-wound resistor with a 10-W rating may be used. A replacement resistor should never have a lower-than-specified wattage rating.

Replacement capacitors should also have the same value rating as the original as well as the same voltage-handling capability. If the capacitor is rated at 0.001 μF at 200 V, the replacement having the same 0.001-μF value but with a voltage rating of 500 would still be acceptable except that the higher voltage rating would result in a larger capacitor that may not conveniently occupy the same space that the defective one does.

10-12. BANDPASS AMPLIFIER TESTS

A bandpass amplifier in television color circuitry is often used between a video amplifier and the color signal demodulators. The function of the bandpass amplifier is to strip all blanking pulse and sync-pulse information from the video signal so that these pulse and sync signals do not cause interference when they finally appear on the screen of a monitor or television receiver. Thus, to perform this function, it is necessary to switch the amplifier circuit into periodic nonconduction during the time blanking pulses or sync pulses appear as part of the original video signal.

The basic components making up a bandpass amplifier are shown in Fig. 10-10. The input signal from the video amplifier is applied to the input element (base) of the transistor as shown. A blanking signal is obtained from the horizontal output system of the monitor or receiver or obtained by sampling the horizontal sweep rate of the signal being received. The blanking signal is applied to the junction of the two capacitors, shunting the RF transformer primary L_1. The blanking signal opposes the applied dc voltage signal, and the short-duration blanking pulse thus shuts off the transistor current flow in the emitter-collector circuit for a brief interval. Because the blanking pulse only occurs during the horizontal blanking interval, it shuts off transistor current only during this time. Thus, amplification of the sync and blanking as well as the burst signal riding on the horizontal blanking are effectively eliminated. Thus, the bandpass amplifier only processes video signal information. The output from the bandpass amplifier is applied to the color demodulators or to another amplifier if additional gain is required.

The resistor placed across the L_2 resonant circuit is for broadening the frequency response to enable the circuit to accommodate the wide range of signal frequencies encountered. The circuit is made broadly resonant around the 3.58-MHz region to accommodate the chrominance sideband components. To reduce to a greater degree the amplitude of the luminance signals that are lower in frequency, additional filtering circuits are used.

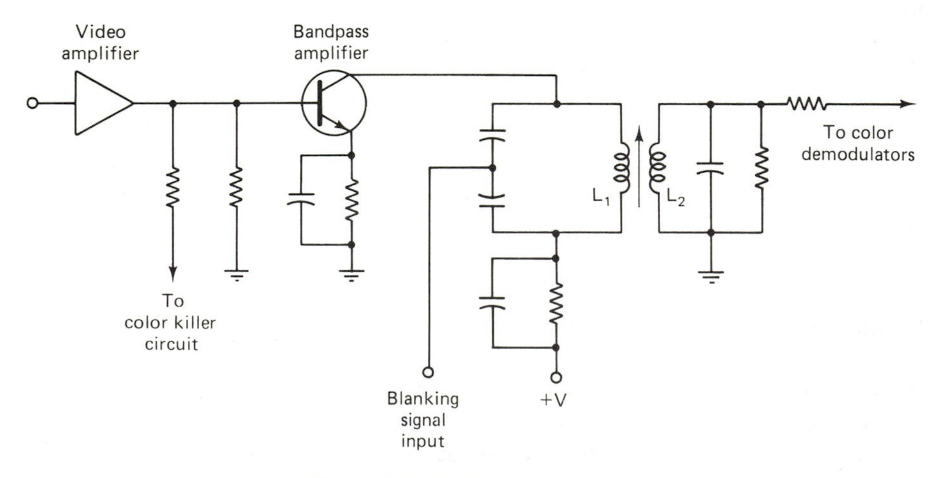

Figure 10-10 Bandpass circuitry.

When it appears that there is some interference with the video signal by the sync and blanking signals, a careful check must be made of the resistors and capacitors. Any resistor that is not approximately the value indicated in the schematic should be replaced. Similarly, all capacitors should be checked for shorted or open-circuit conditions. Capacitors that have one end grounded (the bypass types) drain excessive currents when they are shorted and thus can overload a transistor. If any capacitor is replaced because of a short circuit, a check should also be made of the transistor to make sure it has not been damaged.

Loss of the blanking pulses will cause the circuit to pass undesired signals. An oscilloscope is useful for checking for the presence of the pulses. Since the blanking pulses are applied to the center of the two capacitors, as shown, make sure neither capacitor has become an open circuit because of overload from the high-amplitude blanking pulses. If the service notes from the manufacturer are available, they should be referenced for peak-to-peak values of the pulses as they appear at various sections of the circuit. Also check for open series resistors in the voltage feed lines if a volt-ohmmeter does not indicate proper voltages. Absence of voltage indicates an open series resistor in the voltage feed lines to the base and collector elements of the transistor. This is assuming, of course, that normal voltages appear in other parts of the system. If there is a total low-voltage loss, the power supply and associated components must be checked.

10-13. SUBCARRIER PLL

As discussed in Sec. 1-9, it is necessary to generate a subcarrier signal in any color processing circuitry. Such a signal has been suppressed at the transmitter before broadcasting to conserve space. As mentioned in Sec. 1-9, a reference signal (called a burst) of 3.58 MHz is transmitted on the rear portion of the horizontal blanking. The burst signal is used as the reference to generate the color subcarrier in a crystal-controlled signal generator (oscillator) within the receiver. For proper color rendition, however, the 3.58-MHz oscillator must be locked into precise synchronization with the burst signal so that even a slight phase difference is corrected automatically before color contamination can occur.

To obtain the necessary rigid control of the subcarrier oscillator's frequency, it is monitored by a phase detector that compares the frequency of the oscillator signal with that of the incoming burst signal. If the oscillator signal drifts, a correction voltage is produced that is applied to the input of a reactance control circuit. The latter then alters the frequency of the subcarrier oscillator and thus brings it back to rigid synchronization. The complete circuit is referred to as a phase-locked loop (PLL). A representative system of this type is shown in Fig. 10-11. The general circuits contained in the PLL system are the phase detector utilizing two diodes, a buffer amplifier, a crystal-controlled oscillator, and a reactance control field-effect transistor or a varactor-type reactance control. Note that the circuits form a complete loop from the phase detector to the amplifier, the oscillator, and the reactance control. Thus, both the crystal oscillator signal as well as the 3.58-MHz burst signal are applied to the phase detector. The phase detector senses any difference in frequency or phase between the two signals and rigidly locks in the oscillator's signal with that of the subcarrier burst signal.

Very little trouble should be experienced with the PLL system because the signals handled have very low amplitudes. If the PLL system does not operate properly, it will cause incorrect colors. Before testing the PLL circuit, however, other tests should be made first. If, for instance, the picture has poor or incorrect colors, the tuner should be checked to make sure its bandpass has not been altered. Similarly, if a tint control is present, it should be adjusted for proper flesh tones. A check should also be made on other channels or VCR tapes to make sure the trouble is not caused by a particular station or tape. If these tests indicate normal conditions the PLL should be examined. With the power off, routine resistor and capacitor checks should be made using the ohmmeter scale of a VOM. In the phase detector there must be symmetry between the two sections. That is, the capacitance of C_1 should equal C_2, and the ohmic value of resistor R_1 should also be equal to that of R_2. Very slight variations could, of course, exist, particularly if the tolerance ratings are more than 5%.

Next the two diodes should be checked to make sure no open- or short-circuit condition exists. In checking the diodes, there is no need to disconnect them from the circuit, though consideration must be given to the fact that a diode has a very low resistance in one direction and an extremely high resistance in the other direction. Thus, placing the ohmmeter probes across D_1 does not necessarily indicate a shorted diode if a low ohmic reading is obtained. Reverse the test probes to obtain the high reading. If a very low reading (less than 100 Ω) exists across the diode regardless of the probe interchanges, the diode is defective and must be replaced. The remainder of the routine checks again consist of taking a reading across each capacitor and resistor. The voltages applied to the bottom of resistors R_{10}, R_{11}, R_{13}, and R_{15} should be tested and compared to the voltages indicated in the schematic for this section. An oscilloscope can be placed across resistor R_{15} for an indication of an output signal from the oscillator. Similarly, an oscilloscope can be placed across the input transformer (L_1) of the PLL to make sure a signal from the first amplifier is present. In the absence of signals

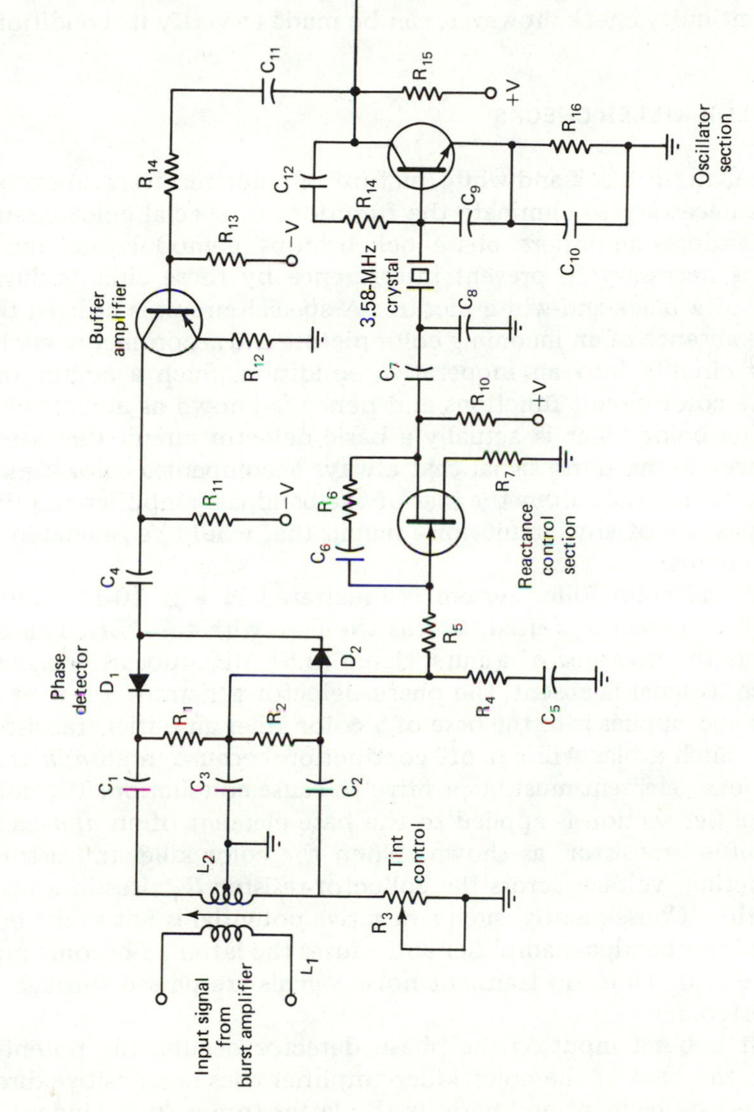

Figure 10-11 PLL circuits.

at the input and output terminals of the oscillator, additional checks must be made of the circuits involved for faulty transistors, resistors, or capacitors. It is unlikely that the input transformer would cause problems because a heavier-gauge wire is used, and volted amplitudes are low. A continuity check, however, can be made to verify its condition.

10-14. COLOR KILLER CHECKS

When receiving a black-and-white picture in color receivers and monitors, it is necessary to eliminate the functions of special color circuits such as bandpass amplifiers, phase-locked loops, demodulators, and so on. This is necessary to prevent interference by these circuits during reception of a black-and-white picture. A special circuit is utilized that senses the absence of an incoming color picture and accordingly switches the color circuits into an inoperative condition. Such a circuit thus "kills" the color circuit functions and hence is known as a *color killer* circuit. The color killer is actually a basic detector circuit that senses the absence of the burst signal that always accompanies color signals. The killer circuit then alters the bias of the bandpass amplifier and thus prevents passage of any circuit-noise signals that would be generated by the color circuit.

A typical color killer system is illustrated in Fig. 10-12. Here a type of phase detector is used, as was the case with the PLL. This circuit detects the presence of a burst signal (3.58-MHz subcarrier signal). When such a signal is absent, the phase detector generates a low or reverse bias and applies it to the base of a color killer amplifier transistor, as shown. Such a bias will cut off conduction because in an *npn* transistor the base element must be positive to cause conduction. The color killer amplifier section is applied to the base element of an *npn* bandpass amplifier transistor, as shown. When the color killer transistor is nonconducting, voltage across the collector-resistor R_7 rises in a negative direction. Consequently, such a negative potential is felt at the base element of the bandpass amplifier and causes the latter to become nonconductive also. Thus, no transient noise signals are passed through to cause interference.

When a burst input to the phase detector occurs, the potential applied to the base of the color killer amplifier rises in a positive direction at the base element and permits the latter to conduct. Under this condition the bandpass amplifier operates normally and processes the video signal that is applied to the base input circuitry, as shown in Fig. 10-12.

Sometimes a small variable resistor (R_{10}) is present on the chassis to permit optimum adjustment of circuit performance. This resistor

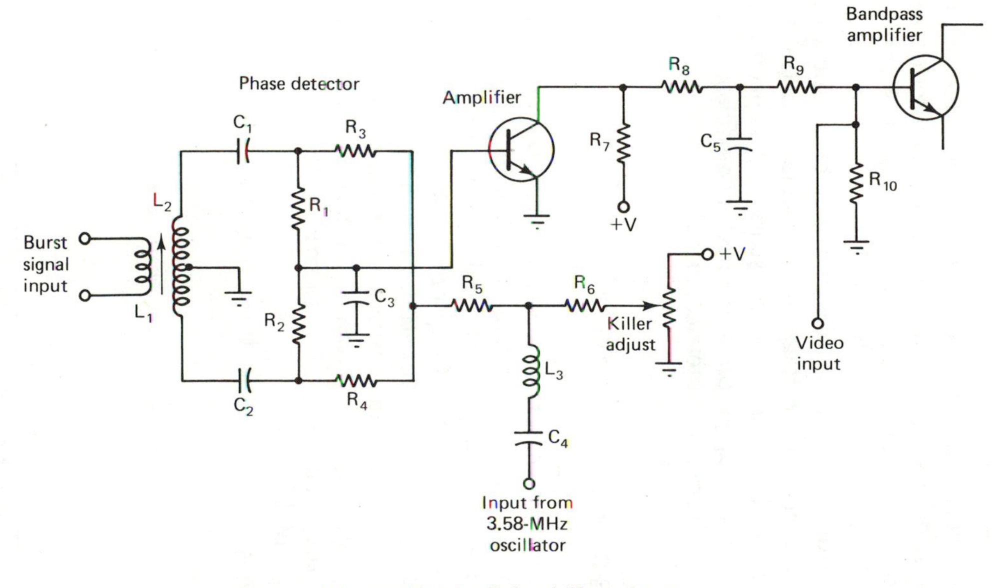

Figure 10-12 Color killer system.

permits compensation for variations in transistor characteristics which may occur as the transistors age. If some interference is present on the viewer screen during black-and-white picture reception, the variable resistor (R_{10}) can be adjusted slightly above and below the normal setting as a check is made of interference-free reception of black-and-white pictures.

Routine tests include a check of the phase detector diodes, transistors, and the resistors and capacitors. An oscilloscope can verify the presence of the burst signal at the input terminal. A reading of the bias on the phase base element of the bandpass amplifier should show a significant amplitude change in bias when a black-and-white picture is being received. If no such voltage change occurs, the transistors must be tested as well as the associated resistors and capacitors. As with all phase detectors, the capacitors and resistors associated with each diode should match to form a symmetrical circuit.

10-15. COLOR DEMODULATOR TESTS

There are several methods for demodulating the chroma signal, and a block diagram of a typical system is shown in Fig. 10-13. The output luminance signal is applied to the color picture tube cathodes or processed as discussed earlier in Chapters 5 and 6 for VCR recording purposes. The chroma signals consisting of the $R - Y$ and $B - Y$ signals are demodulated and mixed in a resistive matrix network to produce the missing $G - Y$ signal. These signals can be processed additionally for recording, or they can be applied directly to the respective control-grid elements of the picture tube.

Sometimes a TV manufacturer may use demodulators termed X

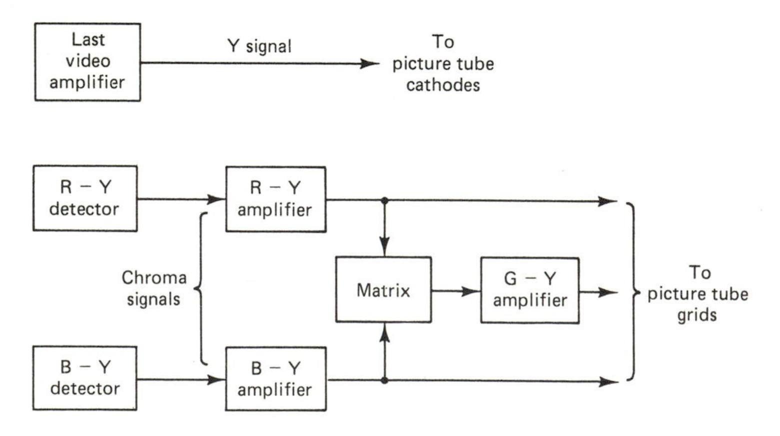

Figure 10-13 Chroma demodulators.

and Z, as shown in block diagram form in Fig. 10-14. The letters X and Z are arbitrary designations for particular phase relationships as set by the manufacturer. Generally the demodulation axes are 105° apart (see Fig. 7-24). As shown in Fig. 10-14, the $G - Y$ signal is obtained by applying the X and Z demodulator outputs to respective amplifiers and then using matrix mixing to obtain the $G - Y$ signals, as was the case with the $R - Y$ and $G - Y$ signals shown in Fig. 10-13. A representative transistorized schematic of the X and Z demodulators as well as an automatic tint control system was shown earlier in Fig. 10-9. There the chroma signal input was applied to the base element of both the Z and X demodulators simultaneously. Similarly, the 3.58-MHz input signal from the phase-locked loop subcarrier generator was applied to the emitter elements, as shown. Thus, the subcarrier signal has been remixed with the color signal sidebands. Now normal detection occurs in the collector element, and the signals are converted to representive $B - Y$ and $R - Y$ signals.

When circuit functions deteriorate, the tests and checks of components and transistors must be undertaken as discussed earlier in this chapter. The oscilloscope is useful to check for the presence of RF signals at the input and output sections. In testing color circuit performance, the vectorscope, discussed in Chapter 7, is also useful. Service notes for a particular unit are always of value for finding specific part values and ascertaining the complete circuit sequences and interconnections. Although defective sections and specific parts can be localized by checks and signal-tracing procedures, a reference to the schematic for a particular device expedites repairs and thus saves time.

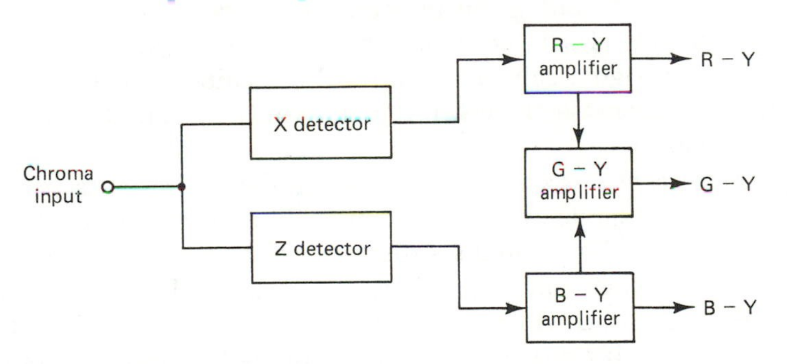

Figure 10-14 *X-Z* demodulator system.

10-16. DEMODULATOR OUTPUT SYSTEM

To drive the picture tube elements, the demodulated video signals usually require additional amplification in similar fashion to that shown in Fig. 10-15. Here is shown the production of the actual three primary

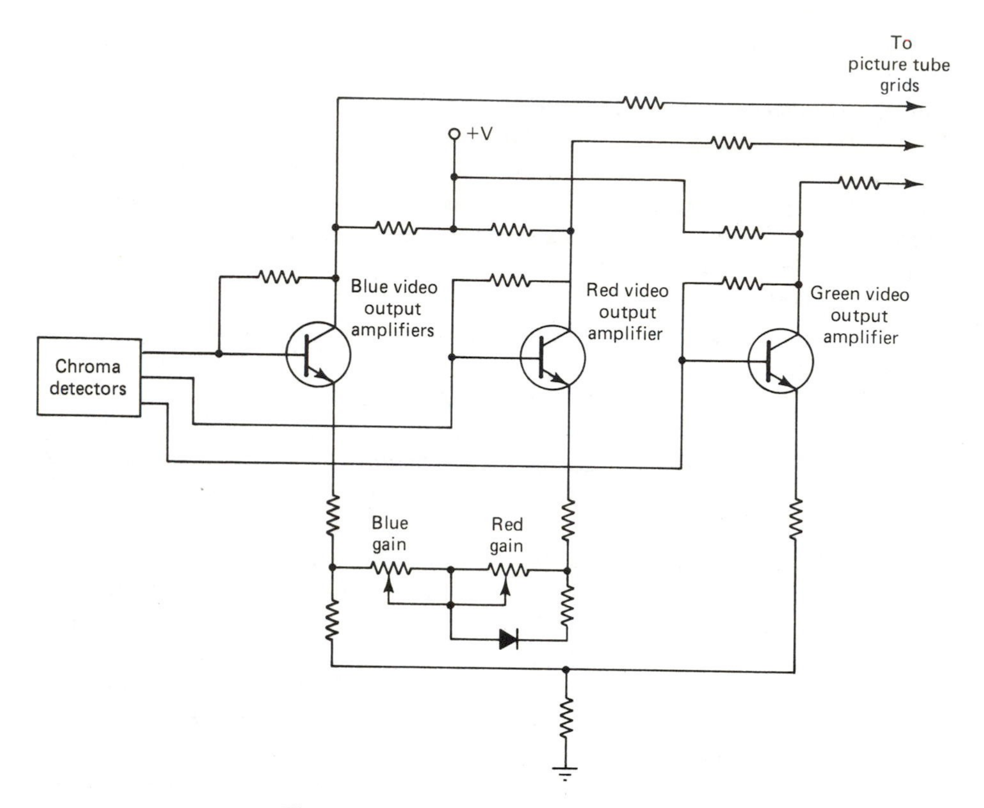

Figure 10-15 Chroma output amplifiers.

colors of blue, red, and green in the *additive color system*. Note that there is a blue signal gain control and a red signal gain control. To increase the green signal gain, the blue and red gain controls are reduced. Conversely, to decrease the green signal gain, the red and blue gain controls are increased.

Spark gap sections are often used in modern sets to discharge accumulated voltage charges from the circuit. When the receiver is new, the high static potential accumulates discharges at more frequent intervals than is the case when operating characteristics level off. Thus, an occasional snapping sound will be heard when the set is used, but this is a normal condition. If a snapping sound occurs with picture deterioration or blackout, the high voltage may be arcing at points other than spark gaps. In such an instance the low-resistance leakage points must be found and the wires and parts separated as much as possible to eliminate high-voltage leakage. Viewing the interior of the receiver during operation in a dark room may give a visual indication of the location of the spark.

Sometimes separate controls are used in the picture tube screen elements, as shown in Fig. 10-16. These are for adjustment of final

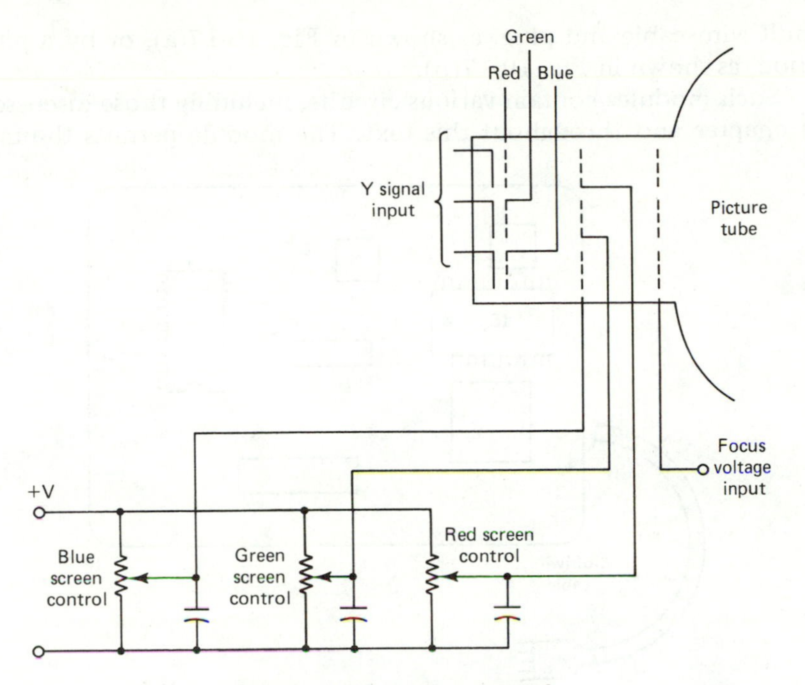

Figure 10-16 Screen controls.

brightness levels of individual colors appearing on the tube face. Usually the three screen controls are available on the rear of the monitor or TV receiver. A service switch cuts off vertical gain and leaves a bright horizontal trace line. If all three color levels are correct, the line should appear white. Two controls can be reduced to check for a particular color gain. If the color for a particular control is of low brightness, the control may be defective or the amplifier circuit malfunctioning. If one color is absent, check the amplifier circuitry and component parts. If a scope shows a signal that can be raised to the same level as the others, it may indicate a defective picture tube with one section burned out. Make sure, however, that a normal signal level is applied to the input line to the picture tube. Also shut off power and use an ohmmeter to verify continuity of the wire in the connecting cable from the circuits to the picture tube.

10-17. CHECKING MODULES

Video cassette recorder circuits, as with other electronic units, not only use integrated circuits (ICs) but usually include the latter in a section of a printed-circuit board called a *module*. Often several such modules are stacked to conserve space. These modules are interconnected either by a multiwire cable and plug, as shown in Fig. 10-17(a), or by a plug-in section, as shown in Fig. 10-17(b).

Such modules contain various circuits, including those discussed in this chapter and throughout this text. The module permits the manu-

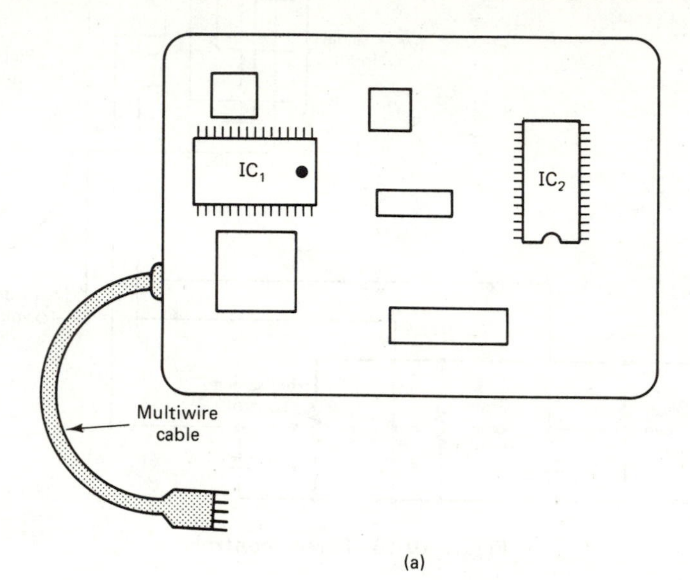

(a)

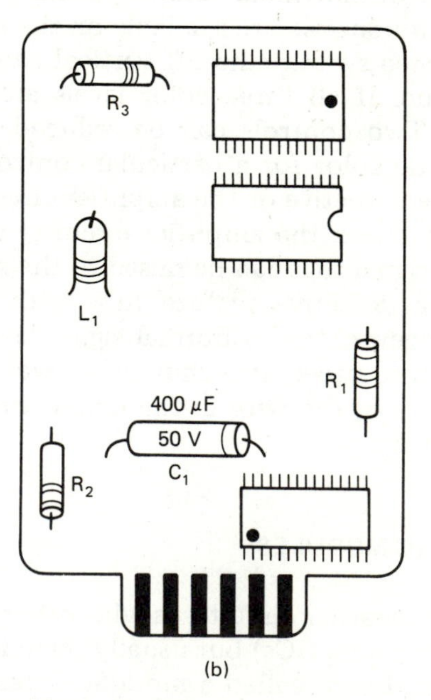

(b)

Figure 10-17 Module connections.

facturer to include variable inductors and trimming capacitors for tuning purposes. In addition the modules contain resistors that must carry more current than can be tolerated by a particular integrated circuit. Similarly, the larger-value (and hence larger-size) filter capacitors are present on the module. Usually a module contains that group of circuits (both wired and formed of ICs) that progress in succession or have similar functions that require interconnections.

When trouble symptoms point to a particular module, the location of the faulty component or IC is facilitated because the immediate circuits associated with the defective one are readily accessible for testing. If continuity checks are needed (where circuit signals and voltages are not required), the VCR is shut off, and a particular module can be unplugged and serviced on the workbench for greater convenience. The general procedures follow other servicing factors discussed throughout this text. Components on the module are checked for continuity, shorted conditions, or off values.

The integrated circuits, if defective, usually alter the amplitude of the voltage applied to the terminals. If the voltage-feed line encounters an open circuit, for instance, the voltage would tend to rise slightly unless the power supply had excellent regulations (the ability to hold the voltage at a fairly even level despite power drain changes). If a partial short exists, it would reduce any applied voltages below normal. Even with a power supply designed for good regulation, the partial short causes a voltage drop below normal. With the partial short consuming abnormal power and raising the current levels, larger voltage drops occur across any series resistors and hence lower the voltage at the area of the defect, causing the partial short.

A short circuit can overheat resistors that are used to drop voltages to the needed value. The shorted IC section, for instance, draws an excessive amount of current from the power supply and overheats or burns out series feed resistors. Before IC replacement is attempted, however, component values and voltage readings should be checked thoroughly. A shorted bypass capacitor, for instance, could draw an excessive amount of current through a voltage feed resistor and cause the latter to overheat. If all checks indicate that the IC is defective, it must be replaced with an exact duplicate in the manner discussed earlier in Sec. 10-4 and illustrated in Fig. 10-3.

A representative module showing feed resistors and bypass capacitors with a single IC is shown in Fig. 10-18. Here the IC has numbered terminals, and the first number position is identified by either a notch in the IC or a colored dot. The signal and voltage feed lines in and out of the IC terminals end at the edge of the module. As mentioned earlier, these can all be connected to a multiwire cable or to plug-in prongs. The common negative line is usually identified by the ground symbol in

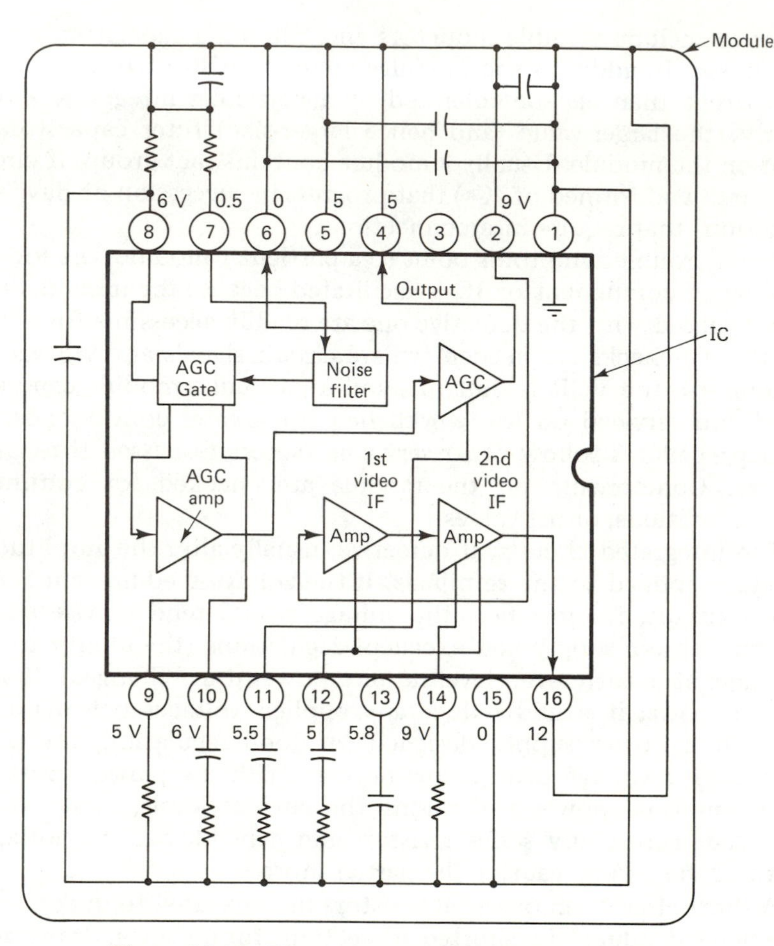

Figure 10-18 IC and associated module components.

standard circuit schematics. The multiple interconnections between the circuits within the IC are not shown in complete form. In most instances only the input and output lines to specific circuits are given. Thus, this particular IC contains the first and second video IF amplifiers. The automatic gain control (AGC) sections include the gate and the input and output amplifiers plus a special noise-filter section. Note the voltage feed lines to terminals 2, 5, and 8. Note that each of these is bypassed to the ground terminal. There are the capacitors which could alter the voltage appearing at these terminals if the capacitors become partially or completely shorted. With a capacitor short circuit there would be maximum current flow in a series resistor, causing it to overheat and often burn out. Thus, if an overheated resistor or an open-circuit resistor is found, the associated capacitor should be checked. If the capacitor checks out all right but the resistor still overheats, it should be disconnected from the IC terminal and the voltage read at the resistor. If it is near normal, shut off voltage at the VCR and check for continuity between the IC terminal and ground. If the resistance shows a shorted condition, the IC must then be replaced. When a series capacitor exists between the IC terminal and a resistor such as at terminals 10, 11, and 12, any dc voltage read at the IC teminal would be blocked from the resistor (by the capacitor).

Often several ICs are present in a single module. As an example, the IC shown in Fig. 10-19 could also be included in the module shown in Fig. 10-18. These circuits receive their signals from the ones shown in Fig. 10-18. Thus, in Fig. 10-19 the video detector is present and receives an input signal from the third video IF amplifier, as shown. In turn, the first video amplifier follows the video detector, and the output video amplifier is also included in this IC. In addition the sync separator is included here because it obtains its signal from the video amplifier section and transfers it to the sweep circuit. In addition, a noise filter is also included to eliminate transient pulses that could interfere with the signals being processed. Interconnections between the two ICs would consist of coupling capacitors or tuned transformers, depending on the particular circuits involved. Also included are voltage-feed resistors, bypass capacitors, and inductors.

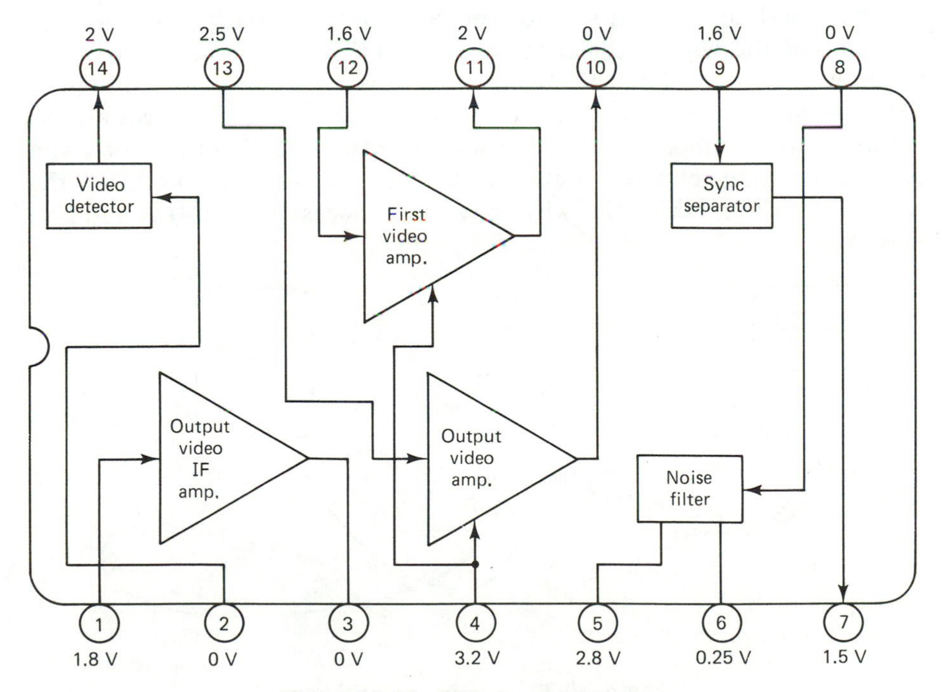

Figure 10-19 Video output sections.

The tests and checks utilized for the module of Fig. 10-18 also apply to the IC shown in Fig. 10-19. Loss of the video signal may be caused by a defect in either the first or second video amplifier IC. The output from the first video amplifier can be observed by an oscilloscope at IC terminal 11 in Fig. 10-19. The output from the second video amplifier can be tested at terminal 10. Failure of the signal to appear here (on an oscilloscope) could indicate a faulty IC. As a precautionary measure, however, the components attached to the output terminal (such as resistors and capacitors) should be checked to make sure these are not shorting the signal to ground. As mentioned for the module, a short circuit could be caused by a defective bypass capacitor.

10-18. REMOTE CONTROL UNIT TESTS

Remote control devices for VCR units are of two major types: the wired and the wireless. A typical wired type is shown in Fig. 10-20(a). When plugged into the VCR, it permits station selection and power on and off and usually includes additional functions such as record, play, fast forward, and reverse. The versatility of such devices depends on the manufacture and design. When the wired types malfunction, the problem often is caused by an open circuit in one of the lines of the connecting cable. To verify this, disconnect the plug from the VCR, open the control box, and check each wire of the cable for continuity between the entrance of the wire to the control box and the exit at the output plug, as shown in Fig. 10-20(b). Such multiwire cables usually consist of colored wires for easy identification of a wire from the control box to the plug. When checking for continuity between the control box and the output plug, clip the meter probes to the terminals and flex the cable. Often a break in the wire may make occasional contact and give false readings.

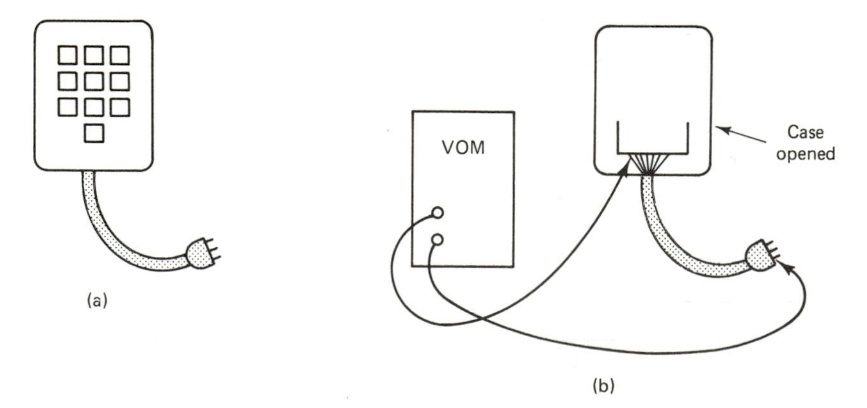

Figure 10-20 Remote control box.

If the cable checks out all right but the control box will not function, check the socket into which the cord plug is inserted. With the VCR disconnected from the power source, remove the outer cover and inspect the wire going to the remote control plug. If all these check out all right, the trouble could be in either the control box circuitry or the control signal sensing circuitry of the VCR, and these must be checked. Since such units differ considerably among the various manufacturers, it is advisable to use the manufacturer's schematic when making repairs on the remote control unit. Often the remote control circuitry is not defective, but one of the push buttons is not making good contact. Thus, the remote control box must be opened and the switches checked carefully for good contact. Again, with the unit removed from the socket, the continuity of each switch within the unit can be checked with an ohmmeter.

The wireless-type remote controls generate a signal either of the microwave type or the infrared. In either case, best results are thus obtained when the transmitting unit of the remote control is pointed toward the VCR. The transducer output port is at one end, as shown in Fig. 10-21(a). One of the primary reasons for failure of wireless remote

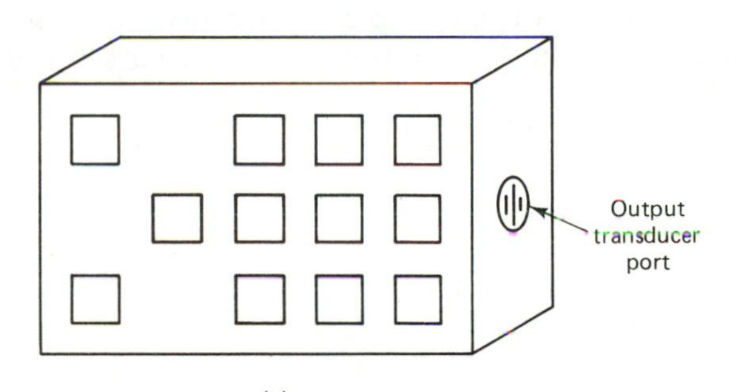

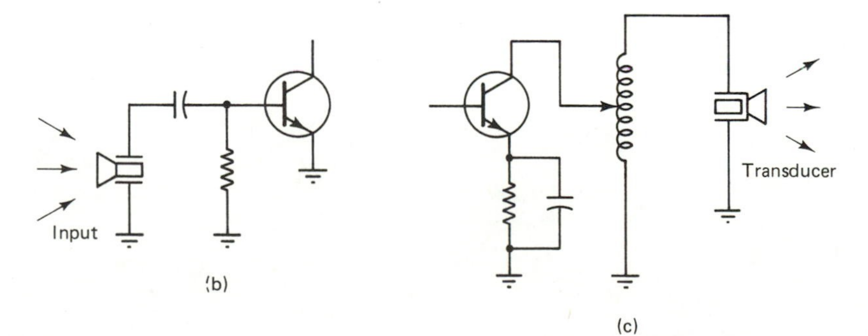

Figure 10-21 Wireless remote control.

control units is the weakening of the batteries in the unit. Since the circuits in the control box contain an oscillator signal generator and amplifier, the battery drain is significant, and for long continuous use of the unit alkaline batteries should be used. When any troubles occur, the batteries should be checked under actual load conditions. That is, the voltmeter test probes should be placed across the batteries and the control box select buttons depressed. Thus, the circuits are initiated and draw current. If the batteries are weak, the voltage will drop perceptibly.

Even with fresh batteries problems sometimes occur because the battery-holding spring or receptacles are not making good contact. Make sure such springs and contact points are clean. (Use a file or sandpaper to remove oxidation.) If a foam-rubber cushion pad is present to press the batteries into their holders, make sure the cushion still has sufficient resiliency to hold the batteries firmly in place.

On rare occasions a faulty remote control operation may be caused by either a defective transmitter unit, as shown in Fig. 10-21(b), or the receiving transducer, as shown in Fig. 10-21(c). Thus, if new batteries do not restore operation, the input and output transducers should be checked. The voltages and signals developed within the remote control unit usually are insufficient to cause burnout of components. The latter, however, can be checked on the printed-circuit board within the unit.

11

Pinpointing Common VCR Troubles

11-1. LOCALIZATION FACTORS

One of the major tasks in troubleshooting and repairing any electronic equipment is finding the exact spot where a defect exists. In some instances quick-check procedures are sufficient to localize the problem, as was discussed in Sec. 2-4 regarding cable-linkage troubles. In other instances the procedure may involve using special test equipment and resorting to signal-tracing procedures and signal testing, as covered in Chapters 9 and 10. On occasion an open circuit or a short circuit may be present, and localization could involve quick-check procedures or more extensive use of test equipment, as discussed in Chapter 7. Thus, the various aspects of testing, checking, and troubleshooting VCRs and monitors are covered in the various chapters in this text. As a convenience, however, much time can be saved and problem localization expedited if a quick-reference listing is available identifying and associating a particular symptom with the probable cause and also indicating the specific chapter or section number giving troubleshooting and repair particulars. Thus, Table 11-1 comprises the balance of this chapter and represents a comprehensive index to the specific chapter or section of this book for a particular trouble symptom.

TABLE 11-1 Master Index to Common VCR Troubles

Symptoms	Probable Cause	Servicing Reference
Abnormal horizontal expansion or compression	TV receiver's horizontal gain or linearity controls misadjusted	Sec. 7-7
Abnormal vertical expansion or compression	TV receiver's vertical gain or linearity controls misadjusted	Sec. 7-7
Automatic tape-loading function faulty	Loading motor inoperative	Sec. 5-10
	Loading motor belt broken or loose	Sec. 5-10
	Loading guide sections bent or broken	Sec. 5-10
Cross-talk interference	Comb filter in VCR defective	Sec. 9-8
Extreme variations of picture contrast	Faults in VCR or TV receiver AGC	Secs. 4-9, 10-7
Gradual picture deterioration	Component aging, effects of humidity, drifts off tuning	Sec. 10-10
Horizontal hum bar in picture and hum in audio	In VCR or receiver: filter capacitor in power supply	Sec. 8-1
	Defective rectifier diode	Sec. 9-7
Incorrect colors	Misadjusted color controls in receiver or monitor	Sec. 1-9
	Mistuned sections	Sec. 7-7
	Misadjusted conversion controls:	
	Defects in ACC and ATC circuits	Sec. 10-9
	Defect in subcarrier PLL	Sec. 10-13
	Defective color demodulator stage	Secs. 10-15, 10-16
Intermittent reception	Interconnecting cable faults	Sec. 2-4
No picture (but roster present), sound normal	One or more video amplifiers defective	Sec. 9-7
	Contrast control defective	Sec. 10-5
No picture or sound but raster present	Interconnecting cable faults	Sec. 2-4
	Fuse open	Sec. 8-6
	On-off switch defective	Fig. 8-7
	Defective tuner or a defect in any stage from the tuner to the sound take-off	Secs. 10-3 to 10-9
	A defective transistor or part in *both* sound and video sections	Secs. 10-3 to 10-9
	An open-circuit condition in the cable or line feeding the tuner	Secs. 10-3 to 10-9

TABLE 11-1 Master Index to Common VCR Troubles (continued)

Symptoms	Probable Cause	Servicing Reference
No sound output, picture normal	Defective audio amplifier stages or speaker	Sec. 10-2
One channel in stereo inoperative or faulty	Cable linkage	Sec. 2-5
	Defective plugs	Sec. 2-5
	Defective audio head	Secs. 5-3, 6-3, 7-5
One stereo channel sound missing, picture normal	Defective audio amplifier, faulty speaker, or connections to speaker	Secs. 2-5, 12-2
Pause control inoperative	VCR control circuit or relay faulty	Secs. 4-11, 12-6
Picture blooms (expands to excess vertically and horizontally)	High-voltage supply in TV receiver defective and voltage below normal	Sec. 8-9
Picture information crowded at either side	Incorrect setting of horizontal linearity control	Sec. 7-7
Picture information crowded at either top or bottom	Incorrect setting of vertical linearity control	Sec. 7-7
Picture interference	VCR cover missing	Sec. 1-4
	Radiated noise	Sec. 1-6
	Improper cable linkage:	
	Misadjusted adjacent-channel traps	Secs. 1-10, 9-5
	Sound IF (41.25-MHz) trap misadjusted	Sec. 1-8
	Defective bandpass amplifier	Sec. 10-12
Picture present but no colors, sound normal	Defective color killer circuit	Sec. 10-14
	Chroma circuits in VCR or receiver defective	Secs. 10-15 to 10-18
Poor picture and sound quality	VCR heads need cleaning and/or demagnetization	Secs. 3-1, 3-5
	Worn tape	Sec. 3-4
	Tracking control adjustment needed	Secs. 5-5, 6-5
	Circuit distortion	Sec. 7-5
Poor picture detail	Misaligned tuner of IF stage in VCR or receiver	Secs. 1-2, 1-3
	Loss of interlace in receiver	Sec. 1-6
	TV receiver focus control needs resetting	Sec. 1-6
	Defective peaking coils	Sec. 9-6
	Faulty tape	Sec. 9-6

TABLE 11-1 Master Index to Common VCR Troubles (continued)

Symptoms	Probable Cause	Servicing Reference
Screen dark, sound normal	Loss of high voltage or defective picture tube	Secs. 8-9 to 8-11
	Brilliancy and contrast controls set too low	Secs. 8-9 to 8-11
	Defective controls	Secs. 8-9 to 8-11
Sound output has noticeable background noise level, picture normal	Defective noise-reduction circuitry (deemphasis and Dolby)	Sec. 12-5
Stop function inoperative	VCR circuits or relay faults	Secs. 4-8, 12-6
Streaks in picture	Loose or defective connections between VCR and TV receiver or monitor	Sec. 1-10
	Worn recording tape	Sec. 1-10
	Heads need cleaning	Sec. 1-10
Uneven or erratic playback (or recording)	Loose belts	Secs. 3-7, 3-8
	Guideposts dirty, rollers in need of oil	Secs. 3-7, 3-8
	Servo control circuit defective	Secs. 5-4, 6-4
Vertical interference bars on TV screen	Caused by presence of signal interference having a frequency higher than 15,734 Hz	Sec. 9-7
Weak picture but normal sound	Radiation losses	Sec. 1-4
	Signal losses in chassis	Sec. 1-4
	Contrast and/or brilliancy controls misadjusted	Sec. 1-10
Weak picture and sound	Cable and VCR input impedance not matched	Sec. 2-1
	Power supply defective	Secs. 8-1, 8-3, 8-4
	VCR or receiver tuner needs tracking	Secs. 9-2, 10-6
	VCR or receiver IF stages need aligning	Sec. 9-4
	AFT system defective or inoperative	Sec. 10-8

12

Miscellaneous VCR-Related Factors

12-1. COMPUTER USAGE FOR TOPIC TITLES

When a home computer is connected to a television receiver, the latter serves as a monitor for computer programs by displaying appropriate data on the screen. Since such a display is produced by a standard video signal input from the computer, the latter can be utilized to print titles or captions to VCR-recorded material. Thus, appropriate designations can be recorded and displayed during playback of a tape for identification of the incident or the participants, and so on. To perform this, the title is typed into the computer and displayed on the screen. Once the appropriate title has been typed, displayed, and corrected, the output from the computer is then fed to the VCR and recorded. The length in time of the display is then determined by how long you maintain the image before shutting off the VCR recording mode.

An alternative method is to feed the computer output into the VCR, setting the VCR input to either channel 3 or 4, depending on the computer output setting. The desired title can then be typed on the computer keyboard, and the VCR monitor or receiver acts as a direct display. (The VCR is switched to the recording display mode normally used in the recording of a TV channel while viewing it.)

If the computer connects to a monitor that is part of the computer keyboard, the video input to the monitor may not be compatible with the VCR because a higher-resolution nonstandard picture may be used in the computer and the monitor. In such an instance the field and frame rates of the signals may have been increased to present a much higher-resolution display.

When displaying titles on the screen and recording them on the VCR, satisfactory results will only be obtained when the recording referred to by the title or caption follows directly after the title has been recorded. It is difficult to dub in a title on a tape that already contains recorded material. If a space has been left on the tape for the inclusion of a title, there will be a blank spot on the tape for a short distance following the recording of the title. This comes about because the erase head that precedes the recording head will have erased a short portion of the tape regardless of when the dubbed-in recording is stopped. If sufficient space has not been left in the recorded tape, some of the prerecorded material will be erased. If enough space has been left, a blank screen period will exist between the ending of the title or caption and the beginning of the related prerecorded material. This is not a fault that can be corrected but is a part of the design.

With modern color computers the title displays can be multicolored, and some graphics can be utilized for novelty effect. When a video camera is utilized with the VCR, there is no problem with a blank area in the tape since the camera can be utilized for taking a picture of a title, and then in a continuous recording process the event in question can be photographed and recorded.

12-2. STEREO AND HIGH-FIDELITY

In Chapters 5 and 6 (Secs. 5-3 and 6-2) the high-fidelity and stereo capabilities in modern VHS and Beta systems were discussed. As pointed out, although high-fidelity tapes can be played on older machines, the resultant sound quality remains inferior when only a single-track segment was used for audio recording purposes. Both the Beta and VHS systems maintain compatibility for their own tapes. Thus, a VHS tape, whether recorded on the old machines or the new, can be played on either the old or new VHS unit. This is also true for Beta machines playing old or new Beta tapes. It is only when the VCR has stereo and high-fidelity capability built in that full advantage can be obtained from newly recorded tapes. One other factor is important, however, and that is that the recorded high-fidelity stereo sound must be played through a high-quality stereo amplifier feeding a good stereo speaker system.

A mono audio recording can be played on a stereo system, and though both speakers will be in operation, the sound source will appear to be fixed in the central area between the two speakers. Similarly, a stereo recording, when played on a mono hi-fi system, will lose its stereo effect when heard from a single speaker or even with several speakers wired either in series or parallel. A good simulation of stereo can be achieved by using a stereo synthesizer device as available from certain jobbers. As shown in Fig. 12-1, such a unit processes the mono signal applied to its input and produces a left- and right-channel output having dissimilar sound levels for various instruments to imitate a stereo receiver output. The device employs phase shifting and passive filter networks to achieve a sense of sound displacement that gives a good imitation of stereo recording. When such a synthesizer is used, provision should be available for bypassing the unit when an actual stereo tape is being played. Again, best results with such add-on units are obtained only if the amplifier is a high-fidelity type (with good stereo separation) that feeds two or more high-quality speaker systems.

As discussed in Sec. 2-5, a compatible stereo sound system is in effect for VCR units and television receivers. To receive and record such stereo broadcasts of the sound that accompanies the video, it is necessary to have appropriate decoding circuitry in the VCR or television receiver. As an alternate method an adapter can be utilized (available from jobbers) which permits the reception of the multiplex signals that produce the left- and right-channel stereo sounds that accompany the video. With an adapter (or a receiver incorporating the appropriate circuitry) true stereo sound is achieved. The fidelity is limited only by the quality of the amplifying system and speakers used within the receiver. Similarly, if the stereo output from the television is fed to a high-fidelity stereo amplifier, the degree of stereo separation, frequency range, and dynamic audio-level differences will depend strictly on the quality of the equipment utilized. As mentioned in Sec. 2-5, the stereo sound obtained by using equipment designed for its reception is the true stereo sound as recorded in the studio and transmitted over the air. A stereo synthesizer unit, however, has the ability to convert a mono signal from a record, television, or other mono sound source to a *synthetic* stereo sound. When there is no direct comparison made between

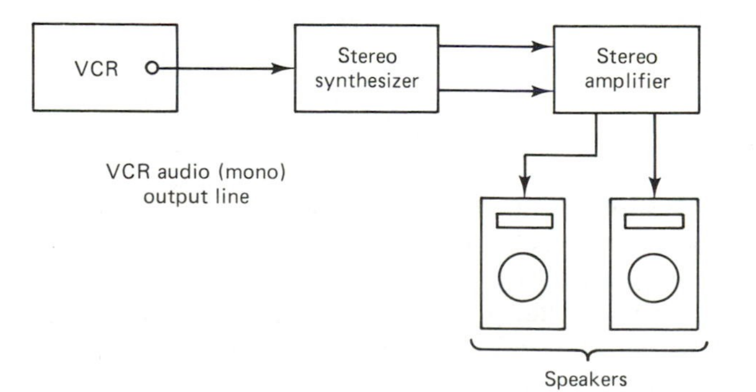

Figure 12-1 Stereo synthesizer usage.

the latter and the true stereo sound, the stereo effect obtained from the synthesizer is sufficiently realistic to serve as a good substitute for the true stereo sound. It is only when, in rare instances, the true stereo sound can be compared directly with the synthesized sound that a marked difference in realism is evident. Interconnections for the synthesizer are shown in Fig. 12-1.

When any troubles occur in either a synthesizer or a true stereo system, the quick-check procedures given in Sec. 2-5 should be undertaken. As with all types of problems that may occur during VCR usage, an isolated instance of poor quality should not be assumed to be a definite fault in the VCR mechanism or its electronic circuitry. The poor quality both in picture and sound could easily be caused by the particular tape utilized. Before extensive tests and troubleshooting procedures are undertaken, another tape (known to be of good or excellent quality) should be used as a test tape to verify whether or not there are circuit malfunctions. Reference should also be made to Fig. 2-7 and the related discussions in Sec. 2-5 for quick-check procedures.

Low-level signals in early RF or audio stages can be handled by small transistors. For higher-amplitude RF signals or audio signals, larger transistors are used, and special precautions are taken to prevent heat buildup. Thus, the output stages of stereo systems contain the larger-type transistors capable of handling signal power ranging from 10 W/channel to well over 100 W/channel, depending on design. In a high-fidelity stereo amplifier, for instance, two transistors are often used in a circuit termed *complementary symmetry*. An example of the latter is illustrated in Fig. 12-2. Note that the output consists of two power transistors feeding a speaker. This schematic represents one channel only, and an identical system would be present for the other channel.

As shown in Fig. 12-2, such a circuit contains no output transformers and thus eliminates the uneven frequency response that may result. Resistors R_1 and R_2 form a voltage divider, and hence the potential at the base input of transistor Q_1 is positive with respect to the potential at the base of Q_2. The latter potentials satisfy the proper bias requirements for the upper *npn* transistor and the lower *pnp* type. The system produces a clean signal output with a minimum of harmonic distortion.

With such power transistors precautions must be taken to satisfy the heat-sink requirements when the transistors are found to be defective and must be replaced. Several heat-sink methods are illustrated in Fig. 12-3. In Fig. 12-3(a) is shown the lower-power transistor that uses a wrap-around metal clamp for dissipating excessive heat. The form of the heat sink is shown in Fig. 12-3(b). In Fig. 12-3(c) is shown a metal flange extending up from the chassis. The transistor is mounted on this flange and transfers its heat to the larger surface area for rapid dissipation.

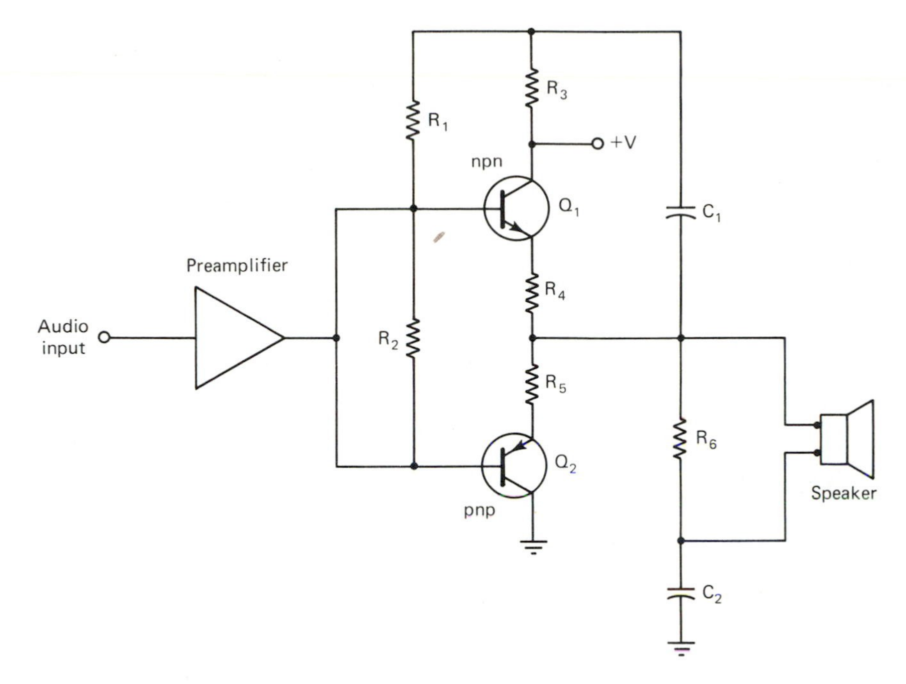

Figure 12-2 Stereo output amplifier (one channel).

The larger-power transistors have a shape as shown in Fig. 12-3(d) and (e). The transistor shown in Fig. 12-3(d) is insulated from the metal chassis by a mica gasket. When such a transistor is installed, the gasket is coated with a heat-conducting grease for better heat transfer to the metal chassis. Such transistors cannot be mounted without the heat sink because the outer metal shell form the collector element and would short the system. For the transistor shown in Fig. 12-3(e), a similar gasket may be required depending on the mounting. For a printed-circuit board with a plastic surface none may be needed. For a metal chassis, however, most transistors require the mica gasket. (The plastic sheet mount does not dissipate sufficient heat for the larger-power transistors.) In Fig. 12-3(f) and (g) are the side view and the bottom view of the transistor shown in Fig. 12-3(d).

The exact mounting details of a power transistor are shown in Fig. 12-4. Note that the mounting bolts are also insulated from the metal chassis by using plastic insulators that fit flush with the metal chassis openings. During installation, coat both sides of the mica gasket with thermal-conductive silicone grease (available from jobbers). Once the grease has been applied, thread the bolts through the holes of the transistor and the gasket, as shown. Tighten the bolts securely but do not overtighten to the point where warping may damage the unit. As a precaution a check can be made between the collector shell of the

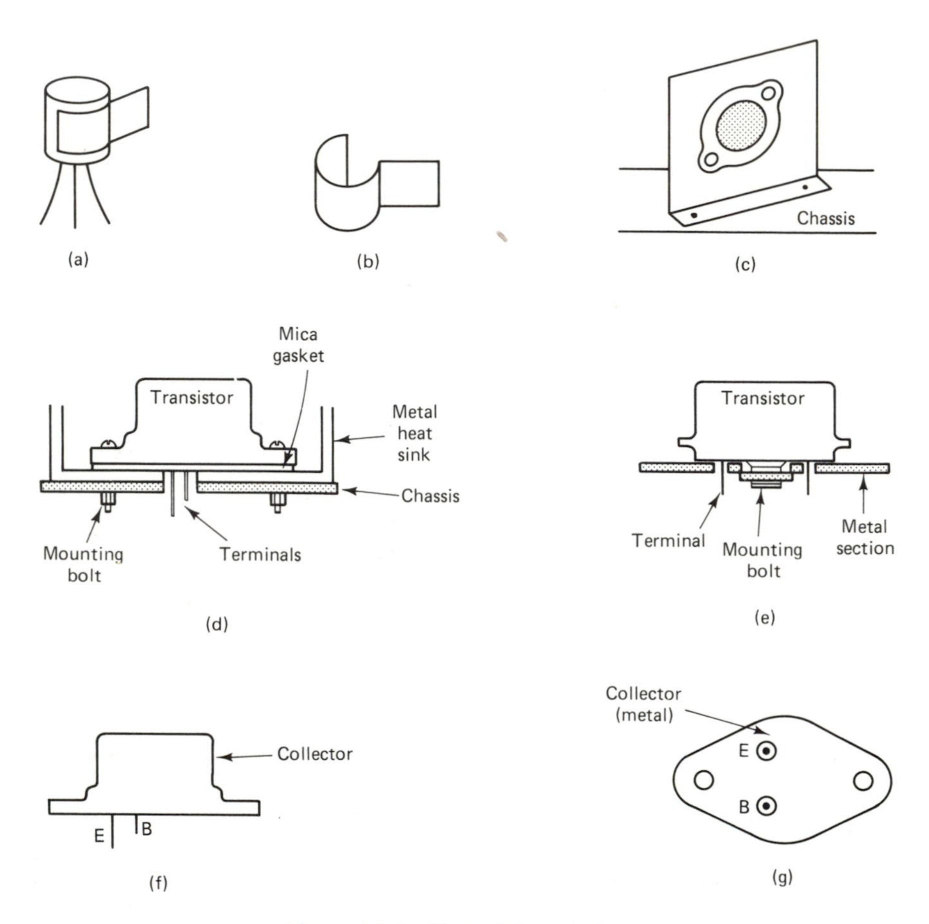

Figure 12-3 Heat-sink variations.

transistor and the metal chassis. An ohmmeter reading between the two points should not show a direct short.

When it is necessary to replace a field-effect transistor (FET), additional precautions must be taken besides heat sinks. Many FET units are subject to damage by static discharge. Some of the FET units are shipped with shorting sections across elements to minimize static damage during shipment. When soldering leads to the FET elements (if no socket is used), it is recommended that the slow-heating soldering iron be used instead of the soldering-gun type. The latter type emits magnetic fields when turned on and off that may cause FET damage. If an exact replacement transistor is not available, consult a transistor cross-reference book (at jobbers) to find an equivalent type with a different identification number.

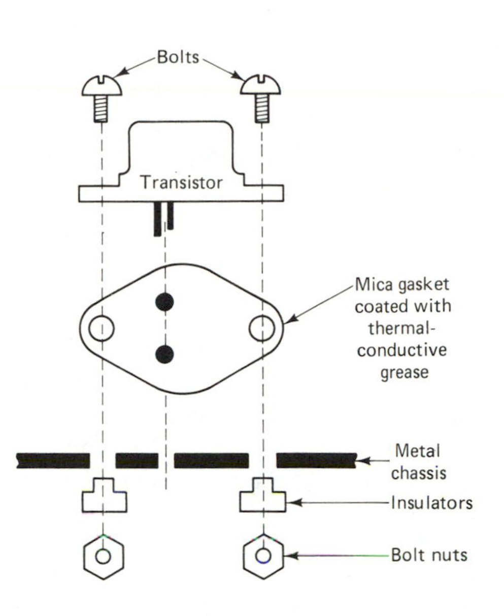

Figure 12-4 Power transistor mounting.

12-3. VIDEO CAMERA DATA

There are a number of commercial cameras available that take pictures and record them on video tape. The state of the art has progressed to the point where the camera is not only small and portable but the VCR can also be carried and the camera attached to it. In other models the camera and recorder units are combined in one package.

If any problems arise in intermittent operation in units where the camera is connected to a recorder with a coaxial cable, the latter should be checked for possible faults. Reference should be made to Chapter 2 regarding problems associated with cable linkages. While in the picture-taking mode the cable can be flexed along its entire length and the recording played back to see whether or not interference patterns appear on the screen. Sometimes the causes of the so-called "glitches" are internal, but they could be caused by intermittent operation in cable linkages.

In the camera-recorder units the resultant recordings on the tape may not achieve true compatibility with the playback unit. In modern machines a compensation (tracking) control is available to alter the playback characteristics slightly to achieve a good match. This is necessary on occasion even though a Beta tape is compatible with any other Beta machine. The same is true for a VHS tape recorded on one VCR but played back on another. Some problems occasionally arise in this area (See Secs. 5-5 and 6-5.)

Inferior playback may not necessarily be caused by any degree of incompatibility between the recorder VCR and the playback. An inferior tape will produce poor results, as will unfamiliarity with the camera controls. The focus, depth of field, lighting, and other similar factors must all be properly recognized and adjusted to obtain good-quality recordings.

12-4. CABLE TV FACTORS

As with television receivers, video cassette recorders can be used on pay cable television. Thus, a VCR may be designated as *cable ready*, as is done with television receivers. Initially, the cable-ready VCRs or receivers were capable of 90 channels, but as the state of the art developed, many VCRs and TV sets were designated as cable ready with 105 channels. Subsequently, this was raised to over 130-channel capabilities. Two factors must be recognized, however, and the primary one is that for reception of pay TV by cable it is still necessary to subscribe to that service. A second factor is that if a VCR is designated as cable ready with 105-station capabilities, it does not imply that this many cable stations will be received in any particular area. The number that are received depends on how many are furnished by the local cable company. The *cable-ready* term simply means that no special switch box is needed, as illustrated in the discussion on cable television frequencies in Appendix B.

If a VCR or television receiver is designated as having a 90-station cable readiness, it may not be sufficient in an area that requires a 105-channel cable readiness. Even though the 90-channel cable-ready VCR may only be receiving 30 channels in that area, the frequency of some of those cable channels may be such that they cannot be received except by a VCR that has a table capability of 105 channels. As detailed in Appendix B, however, a converted unit can be utilized wherein the mid- or superchannel cable stations are converted to UHF frequencies and can thus be received as UHF channels.

Regardless of the number of stations that can be received for a particular cable-ready receiver, it will be unable to decode any cable station that is scrambled. The latter stations require an additional fee for receiving this channel, and the cable company furnishes a decoder to provide for reception.

With older VCR units having only a 12-channel cable capability, the stations that can be tuned in are not necessarily the VHF stations. Instead, of the 12 stations available, some can be UHF. Indicator tabs are usually furnished for visual display of the correct stations which have been selected. With such a VCR, however, the 12-station maximum falls far short of what most cable companies furnish. To take maximum advantage of all the stations furnished by a local cable company, both the VCR and receiver should have a total channel capacity above 130.

12-5. PREEMPHASIS, DEEMPHASIS, AND DOLBY

The sound carrier transmitted simultaneously with the video is frequency modulated. The bandwidth allocation for the sound in television transmission is only 50 kHz instead of 200 kHz as in standard FM radio broadcasting. Consequently, there is some deterioration in the sound quality when the VCR sound signal is recorded only on the narrow 1-mm section of the tape. There is also a higher residual background noise level present in the sound. In FM broadcasting and receiving a special noise-reduction system is employed to minimize circuit noises, transistor noises, and other interfering high-frequency noise signals. Once such noises become an integral part of the FM sound signal, they tend to have a fixed amplitude as well as a frequency span in the upper regions of the audio spectrum.

The characteristics of noise signals are such that they permit the design of special noise-reduction systems. The noise-reduction process is based on the fact that the signal-to-noise ratio can be increased before broadcast to the point where the audio signal is raised in amplitude above the noise level. Hence, by comparison, the background noise remains at a much lower level compared to the average signal. During reception the noise signals as well as the sound signals are decreased in amplitude until the desired sound signal is at its normal level again. Since most circuit noises and background noises exist in the upper audio-frequency regions, the noise-reduction system involves the amplitude increase of the high-frequency audio signals. This procedure at the FM transmitter is referred to as *preemphasis*.

In the United States the Federal Communications Commission establishes regulations regarding preemphasis characteristics, and such regulations are uniform throughout the industry. Thus, the rate of incline of the preemphasis process starts at approximately 400 Hz and rises gradually as the frequency increases. At 1 kHz the increase is 1 db, at 1.5 kHz the increase is near 2 dB, at 2 kHz the increase is almost 3 dB, and at 2.5 kHz the increase is very near to 4 dB. From the latter level upward, the preemphasis rise is substantially linear, reaching 8 dB at 5 kHz and 17 dB at kHz. The basic circuit for preemphasis is simple and consists of a capacitor and resistor, as shown in Fig. 12-5.

For preemphasis the value of C_1 is selected to provide the necessary rise in amplification for the high-frequency audio signal. A capacitor's opposition to the transfer of signals decreases as the frequency

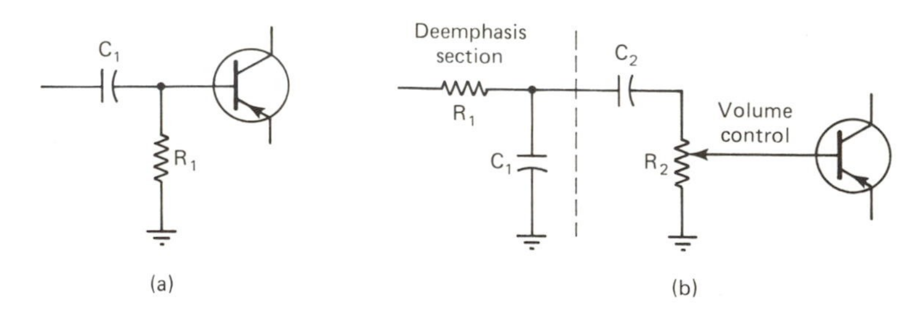

Figure 12-5 Preemphasis and deemphasis.

rises. The opposition offered by a capacitor is termed *capacitive reactance* (X_C). Unlike resistance, which remains substantially fixed regardless of the signal frequency, the reactance of a capacitor is influenced directly by the frequency, as shown by the following equation:

$$\text{reactance } (X_C) = \frac{1}{6.28fc} \tag{12-1}$$

where f = frequency, hertz
c = capacitance, farads

When we multiply capacitance by resistance, we obtain what is known as the *time constant* of a circuit. One time constant represents 63% of capacitor charge voltage. Also, the percentage of capacitor discharge voltage (or change in current) is 37 for one time constant. Thus, if a capacitor is rated at 0.2 μF and the resistor in series with it is 10,000 Ω, the time constant is 0.002 sec, indicating the time at which the capacitor charges to 63% of full value.

For preemphasis the time-constant network is 75 μsec; that is, $RC = 75 \times 10^{-6}$. This value produces best results without unnecessarily increasing the frequency shift of the FM carrier. Generally, an increase in the high-frequency signal amplitude in FM transmission causes the carrier signal to shift frequency (deviate) to a greater extent each side of the resting frequency (the latter is the frequency without modulation). An additional 25-μsec preemphasis is also used on occasion, as discussed later in this section. When the FM signal with the preemphasis arrives at the FM detector, it is necessary to reverse the preemphasis process to restore an even-amplitude frequency response. This process is performed by a *deemphasis* circuit, which, in its basic form, is as shown in Fig. 12-5(b). As illustrated, the deemphasis circuit consists of a series resistor R_1 and a shunting capacitor C_1. The other capacitor (C_2) is the standard intercircuit coupling capacitor, and R_2 is the volume control. The deemphasis circuit follows the FM detector and usually precedes the first audio amplifier. For such a deemphasis circuit the time constant is again 75 μsec so that the existing degree of amplitude increase is decreased proportionately. Hence, higher-frequency signals encounter a decreasing reactance for C_1 and hence are shunted.

Another preemphasis system often used in FM is the Dolby noise reduction. This system was originally adopted because of the widespread usage and success of the Dolby-D noise reduction system in stereo tape recorders. The Dolby system is also useful in reducing background noise in video cassette recorders. The Dolby circuitry is usually encased in a single IC chip and is an integral part of the system circuitry. In the Dolby process during recording, the high-frequency, low-amplitude music signal levels are boosted. Again, the system lifts the noise-prone high-frequency signals above the constant-level background noise. Upon playback, decoding circuits reverse the signal amplitude rise process and decrease the abnormal levels to normal levels. Consequently, the usual tape background noises are reduced considerably. In practice, the reduction of high-frequency noise signals can be as much as −10 db.

The preemphasis process in frequency-modulation systems tends to increase the bandwidth because of the increase in carrier deviation each side of the normal resting frequency. In Europe the preemphasis process uses a less objectionable 50-μsec time constant instead of the 75 μsec used in the United States. For high-fidelity transmission, however, FM stations may utilize compression circuits or automatic peak-limiting devices to prevent excessive bandwidth. With the addition of the Dolby system, the selected preemphasis and deemphasis were reduced to 25 μsec. This is a satisfactory noise-reduction system for FM broadcasting, but it requires a decoder in the receiver to achieve the best noise reduction. If the receiver does not have the proper decoder circuit, the reproduction of high-frequency sounds is altered. When receiver systems are properly equipped, manual selection of the 75- or 25-μsec deemphasis system is available or a built-in Dolby decoder circuit is present.

12-6. RELAY TYPES AND TESTS

Relays are essentially electric switches that are tripped to either the on or off position by the application of voltage to the relay coil. Relays usually contain at least one movable metal section. The end of the latter extends over the iron core of the relay coil, as shown in Fig. 12-6(a). The movable section is identified by (M), and when the coil is energized by the application of the proper voltage, the magnetic field exerts an attraction for the movable switch portion. The latter is thus pulled down and in so doing touches another metal section (4) and thus closes the switch. The relay shown in Fig. 12-6(a) is termed *normally open* because it is closed only when voltage is applied to the coil at terminals 1 and 2.

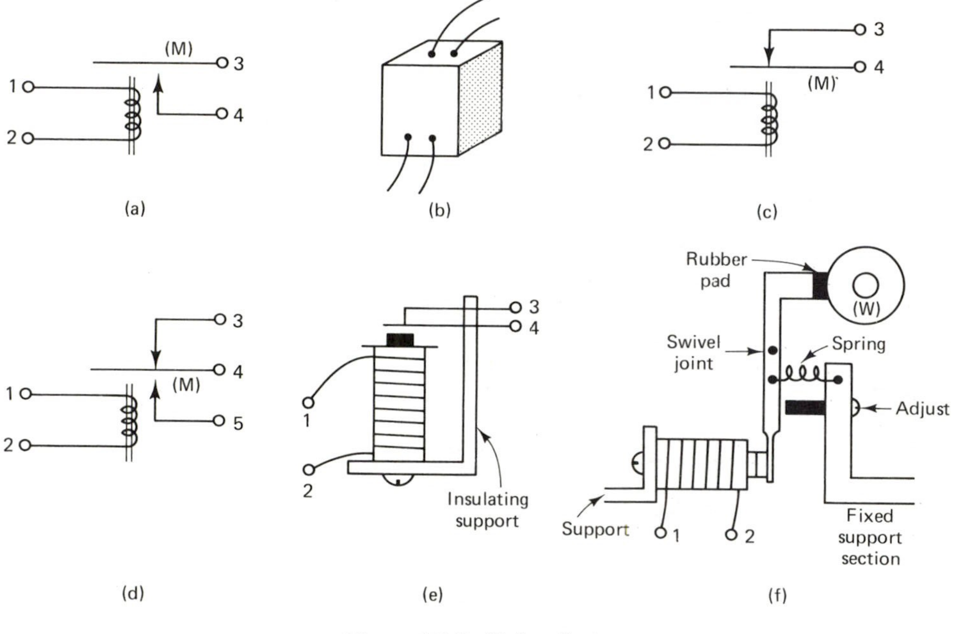

Figure 12-6 Relay factors.

Often a relay is enclosed in a plastic housing, as shown in Fig. 12-6(b). The coil and switch terminals are brought out as shown, and the enclosure protects the relay from dust. The relay shown in Fig. 12-6(c) is a *normally closed* type and opens only when the coil is energized and pulls down the movable (M) arm. At this time the moving portion disengages from the upper terminal and opens the circuit. A dual-type relay featuring both a normally closed and a normally open section is shown in Fig. 12-6(d). When the coil is energized, rod (M) opens the circuit between terminals 3 and 4 and closes the circuit between terminals 4 and 5.

The physical appearance of the relay is shown in Fig. 12-6(e). A coil (also termed a solenoid) contains an iron-core rod that acts as the magnetic core. An insulating support section clamps the coil tightly and also serves as the post supporting the fixed and movable elements of the switch section. Relays such as these can be used to switch motor power on or off as required and can also be used wherever electronic switching (such as by a chip or transistor) is not capable of handling the current involved. Another application of the magnetic principle involved with solenoids is illustrated in Fig. 12-6(f). Here a quick-brake system is used on one of the wheels involved in VCR gear. Under normal play or record conditions the coil is not energized; hence the spring pulls the brake rod away from the wheel. When the coil is energized, it pulls the bottom of the brake rod to it and thus presses the brake pad against the wheel rim. By such methods sudden stops can be achieved.

When a trouble symptom pinpoints a relay as the prime cause, several checks can be made to determine whether the coil or the switch section is at fault. For the arrangement in Fig. 12-6(f), for instance, failure of operation could be caused by an open coil or lack of voltage. Thus, using a VOM on the voltage scale, place the probes across the input lines (1 and 2) to the coil and note if voltage appears when the stop mode is initiated. If not, disconnect one coil lead (either 1 or 2) from the input lines. Shut off power to the VCR and check for continuity in the coil (with the ohmmeter section of the VOM). If the coil is open, it will have to be replaced unless the break is in the soldered connections at the entrance to the relay housing. If voltage and coil check out all right, look for binding at the swivel joint and use a drop of oil here. For switching relays, also check the switching terminals to make sure switches function properly when the coil is energized.

Appendix A

TV Station Allocations

Table A-1 lists VHF television station allocations in the United States. These television stations broadcast public entertainment programs at specific frequencies for each channel as shown. The allocation for each channel is 6-MHz total bandwidth that includes both the video- and audio-modulated carriers. The video portion is amplitude-modulated, and the audio portion is frequency-modulated. The preferred intermediate frequencies (IFs) are 45.75 MHz for the picture carrier and

TABLE A-1 VHF Television Station Allocations (U.S.)

Channel Number	Frequency (MHz)	Video Carrier	Sound Carrier
1	Not used		
2	54–60	55.25	59.75
3	60–66	61.25	65.75
4	66–72	67.25	71.75
5	76–82	77.25	81.75
6	82–88	83.25	87.75
	FM band (88–108 MHz)		
7	174–180	175.25	179.75
8	180–186	181.25	185.75
9	186–182	187.25	191.75
10	192–198	193.25	197.75
11	198–204	199.25	203.75
12	204–210	205.25	209.75
13	210–216	211.25	215.75

41.25 MHz for the sound carrier. These same specifications also apply to the UHF station listing given in Table A-2. Both picture and sound IF signals are heterodyned in the receiver to produce a final sound IF of 4.5 MHz.

Note that the UHF television station allocations are consecutive 6-MHz allocations for each successive channel number from 14 to 83. For VHF, however, there is a break in the frequency sequence between channels 4 and 5. Note that channel 4 ends at 72 MHz but that channel 5 starts at 76 MHz. Similarly, between channels 6 and 7 there is also a break in the frequency sequence since the FM radio broadcast band lies between 88 and 108 MHz.

TABLE A-2 UHF Television Station Allocations (U.S.)

Channel Number	Frequency Range (MHz)	Picture Carrier (MHz)	Sound Carrier (MHz)
14	470-476	471.25	475.75
15	476-482	477.25	481.75
16	482-488	483.25	487.75
17	488-494	489.25	493.75
18	494-500	495.25	499.75
19	500-506	501.25	505.75
20	506-512	507.25	511.75
21	512-518	513.25	517.75
22	518-524	519.25	523.75
23	524-530	525.25	529.75
24	530-536	531.25	535.75
25	536-542	537.25	541.75
26	542-548	543.25	547.75
27	548-554	549.25	553.75
28	554-560	555.25	559.75
29	560-566	561.25	565.75
30	566-572	567.25	571.75
31	572-578	573.25	577.75
32	578-584	579.25	583.75
33	584-590	585.25	589.75
34	590-596	591.25	595.75
35	596-602	597.25	601.75
36	602-608	603.25	607.75
37	608-614	609.25	613.75
38	614-620	615.25	619.75
39	620-626	621.25	625.75
40	626-632	627.25	631.75
41	632-638	633.25	637.75
42	638-644	639.25	643.75
43	644-650	645.25	649.75
44	650-656	651.25	655.75
45	656-662	657.25	661.75
46	662-668	663.25	667.75
47	668-674	669.25	673.75

TABLE A-2 UHF Television Station Allocations (U.S.) continued

Channel Number	Frequency Range (MHz)	Picture Carrier (MHz)	Sound Carrier (MHz)
48	674-680	675.25	679.75
49	680-686	681.25	685.75
50	686-692	687.25	691.75
51	692-698	693.25	697.75
52	698-704	699.25	703.75
53	704-710	705.25	709.75
54	710-716	711.25	715.75
55	716-722	717.25	721.75
56	722-728	723.25	727.75
57	728-734	729.25	733.75
58	734-740	735.25	739.75
59	740-746	741.25	745.75
60	746-752	747.25	751.75
61	752-758	753.25	757.75
62	758-764	759.25	763.75
63	764-770	765.25	769.75
64	770-776	771.25	775.75
65	776-782	777.25	781.75
66	782-788	783.25	787.75
67	788-794	789.25	793.75
68	794-800	795.25	799.75
69	800-806	801.25	805.75
70	806-812	807.25	811.75
71	812-818	813.25	817.75
72	818-824	819.25	823.75
73	824-830	825.25	829.75
74	830-836	831.25	835.75
75	836-842	837.25	841.75
76	842-848	843.25	847.75
77	848-854	849.25	853.75
78	854-860	855.25	859.75
79	860-866	861.25	865.75
80	866-862	867.25	871.75
81	872-878	873.25	877.75
82	878-884	879.25	883.75
83	884-890	885.25	889.75

Appendix B

Cable Television Frequencies

Since cable television uses a direct linkage between the cable company's transmission center and the television receiver obtaining such transmission, the cable company can shift channels around to better suit its needs. Thus, channel 3 may actually be received on the frequency normally used by channel 2. In such an instance the receiver must be tuned to channel 2 in order to receive channel 3. With a VCR, however, the tuning can be altered and the display adjusted accordingly. Either proper tabs can be inserted, as shown in Fig. B-1, or the proper LCD display can be selected in models using this method. When, however, the station selection is made with the receiver, a listing has to be prepared indicating the station that is to be received for a particular channel number appearing on the dial.

Another factor in cable transmission is that the cable company need not transmit UHF stations at the high frequency in which they are transmitted. As signal frequencies are raised, loss factors tend to increase. Thus, the frequencies between channels 6 and 7 consisting of a 20-MHz span can be utilized to receive additional stations. When this is done, however, older receivers would need a box, such as shown in Fig. B-2, or a cable converter, as shown in Fig. B-3. With modern receivers provisions are built in for accepting as many as 135 or more cable channels.

Channels 2 to 13 can be viewed directly on the VHF portion of the receiver, but, as mentioned earlier, the actual channel number may not coincide with the number appearing on the dial. Mid- and superband channels are used by many cable companies. The mid- and superband

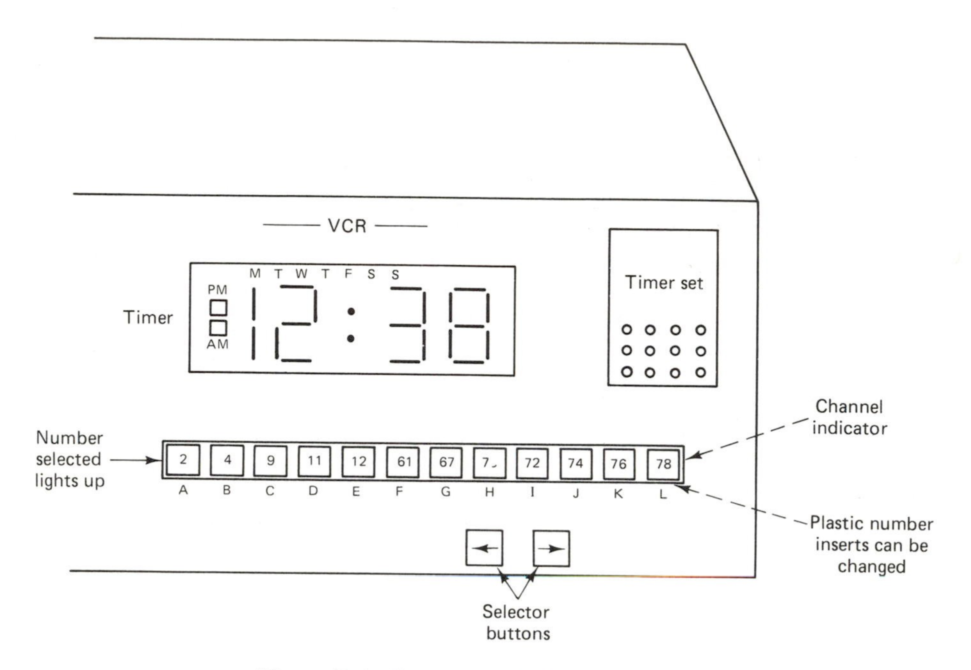

Figure B-1 Tab inserts indicate channels.

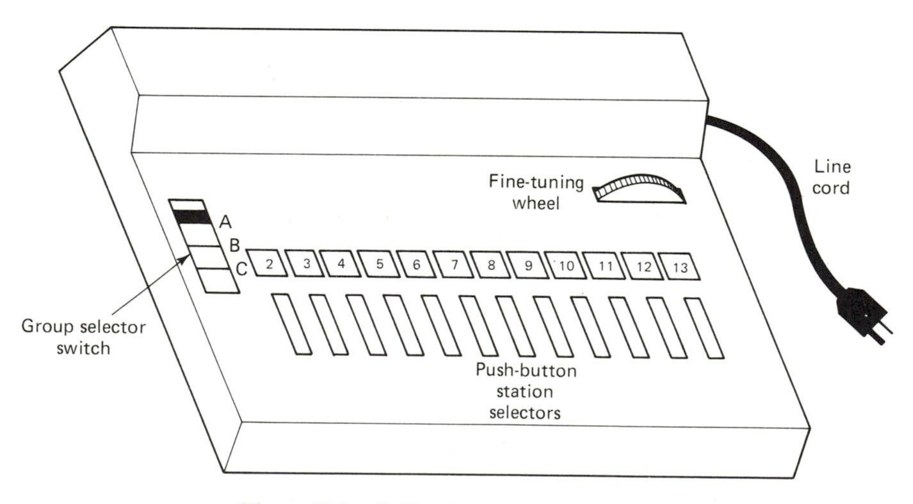

Figure B-2 Cable channel selector unit.

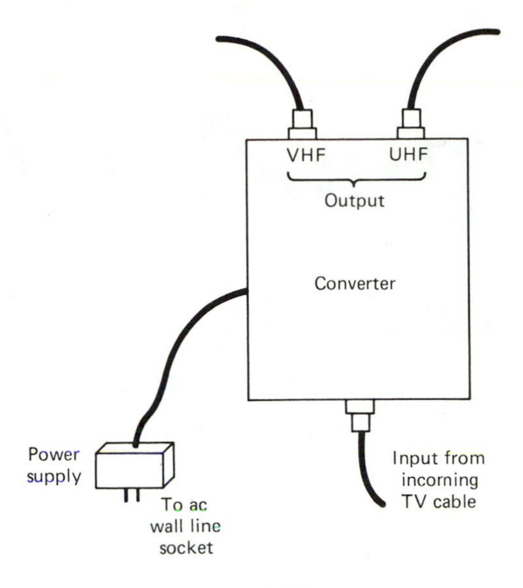

Figure B-3 Cable converter.

channels are converted to UHF frequencies by cable converter units. The UHF frequencies are then available using the UHF tuner of the receiver or the VCR. The VHF channels are passed intact through the converter to the VHF tuner.

Appendix C

TV Standards in Broadcasting

The standards established by the Federal Communications Commission for television are listed in Table C-1. Included are time durations for fields and frames as well as duration of vertical and horizontal sync pulses. Note that the horizontal scan frequency for black-and-white transmission is 15,570 Hz, while that for color reception is 15,734 Hz. Of interest also is the fact that the vertical scan frequency for black and white is the same as the ac in the power mains (60 Hz) but that the vertical scan frequency for color is slightly less (59.94 Hz). In modern circuitry these slight differences are automatically regulated to maintain good vertical and horizontal stability. The proportionate color signal differences are also listed in Table C-1.

TABLE C-1 Television Technical Standards

Function	Data
One horizontal sweep cycle (start of one horizontal line trace to start of next)	63.5 μsec
Horizontal blanking interval	10.16 to 11.4 μsec
Horizontal trace (without blanking time)	53.34 μsec
One frame	33,334 μsec
One field	16,667 μsec
Horizontal sync pulse duration	5.08 to 5.68 μsec
Vertical sync pulse interval (total of six vertical blocks)	190.5 μsec
Vertical blanking interval	833 to 1300 μsec for each field
Vertical scan frequency (B/W)	60 Hz
Vertical scan frequency (color)	59.94 Hz

TABLE C-1 Television Technical Standards (continued)

Function	Data
Horizontal scan frequency (B/W)	15,750 Hz
Horizontal scan frequency (color)	15,734.264 Hz
Total frequency span for an individual station (B/W or color)	6 MHz
Picture carrier is nominally above the lower end of the channel (B/W or color)	1.25 MHz
Aspect ratio of B/W or color picture (picture width versus height)	4.3
Scan lines per frame (B/W or color)	525 (interlaced)
Scan lines per field (B/W or color)	262.5
The frequency-modulated sound carrier is above the picture-carrier frequency (B/W or color)	4.5 MHz
Maximum deviation of sound carrier each side of center frequency	25 kHz
The effective radiated power of the audio expressed as the percent of the peak power of the picture-signal carrier	50 to 70
Color-picture carrier frequency	3.579 MHz (3.58 MHz)
Transmitted burst signal sync	Eight cycles minimum
Brightness (luminance) portion of color transmission, symbol Y	0.59 green, 0.30 red, 0.11 blue
I (in-phase) signal combines portions of $B - Y$ and $R - Y$	$-0.27B$ less the Y signal and $0.74R$ less the Y
The Q (quadrature) signal combines portions of $B - Y$ and $R - Y$	$0.41B$ less the Y and $0.48R$ less the Y
Blue signal combines portions of Y, Q, and I	(Y plus $1.72Q$) less 1.11 of I
Green signal combines portions of Y, Q, and I	(Y less $0.64Q$) less 0.28 of I
Red signal combines portions of Y, Q, and I	(Y plus $0.63Q$) plus 0.96 of I

Appendix D

Frequency Ranges of Letter-Symbol Designations

Earlier in Table A-1 the television broadcast very-high-frequency (VHF) channel listing was given. Included therein was the frequency-modulation radio broadcast band (88–108 MHz). Actually, however, these stations do not utilize the entire VHF spectrum, nor do the UHF stations listed in Table A-2 encompass the entire ultra-high-frequency band. For reference purposes the entire frequency span of all the general frequency designations are given in Table D-1.

TABLE D-1 General Frequency Designations

Designation	Frequency range
VLF (very low frequencies)	3 Hz to 30 kHz
LF (low frequencies)	30 kHz to 300 kHz
MF (medium frequencies)	300 kHz to 3 MHz
HF (high frequencies)	3 MHz to 30 MHz
VHF (very high frequencies)	30 MHz to 300 MHz
UHF (ultrahigh frequencies)	300 MHz to 3000 MHz
SHF (superhigh frequencies)	3 GHz to 30 GHz
EHF (extra-high frequencies)	30 GHz to 300 GHz

Appendix E

Resistor Color Codes

Many types of resistors are encountered in electronic circuitry. In most instances the terminal leads at each end are bent over and inserted into appropriate holes in the printed-circuit board. The leads are then soldered into place and the wire ends that extend beyond the solder point are cut off. The most common type of resistor is the molded composition carbon type illustrated in Fig. E-1(a). This resistor is relatively inexpensive compared to the wire-wound types that carry substantially more currents and have a higher wattage rating. The molded composition type shown is available in a variety of values to meet the requirements in different branches of electronics.

The molded composition carbon resistor is also available in several wattage ratings, the most common of which are the ¼- and the ½-W types. Occasionally a 1- or 2-W value may be used for special circuitry where a smaller value may generate too much heat and be damaged. The film resistor illustrated in Fig. E-1(b) is similar to the molded composition types in appearance.

The accepted method for identifying the ohmic value of a resistor is to imprint bands of color at one end, as shown in Fig. E-1. The particular color code is identified by starting with the first band of color nearest the end of the resistor, as shown. Usually four bands are present on the molded composition resistor and five color bands for the film resistor. The last color band in either resistor indicates the tolerance for the ohmic value indicated by the initial color bands. Thus, if a 200-Ω resistor has a tolerance rating of 10%, the actual value may be anywhere between 180 to 220 Ω.

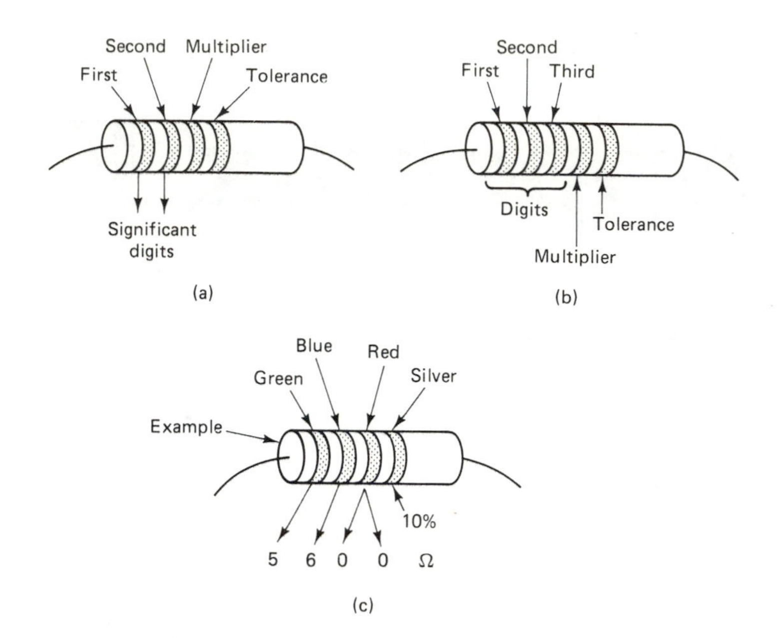

Figure E-1 Resistor color-code bands.

The listing shown in Table E-1 indicates color-code values and tolerances of the resistors shown in Fig. E-1. The abbreviation GMV indicates *guaranteed minimum value*. A tolerance marking of *alternate* (alt.) indicates a coding that may have been used in the past but is now referred to be the *preferred* (pref.) designation.

Although the foregoing data on resistors have been generally ac-

TABLE E-1 Ohmic Values of Color Bands

Color	Digit	Multiplier	Carbon ± Tolerance	Film-Type Tolerance
Black	0	1	20%	0
Brown	1	10	1%	1%
Red	2	100	2%	2%
Orange	3	1000	3%	
Yellow	4	10,000	G MV	
Green	5	100,000	5% (alt.)	0.5%
Blue	6	1,000,000	6%	0.25%
Violet	7	10,000,000	12.5%	0.1%
Gray	8	0.01 (alt.)	30%	0.05%
White	9	0.1 (alt.)	10% (alt.)	
Silver		0.01 (pref.)	10% (pref.)	10%
Gold		0.1 (pref.)	5% (pref.)	5%
No color			20%	

cepted as standard in the industry, deviations in identifying values may be encountered. In some instances the ohmic value is imprinted directly on the resistor, for instance. Where no color coding is present and the resistor has been found to be shorted or open, the schematic for the VCR or monitor will have to be consulted to ascertain the correct value of the replacement resistor. Even when a color coding is present, the actual colors may have been altered if the resistor overheated substantially before breakdown. In such an instance the colors may be difficult to read or inaccurate in their representation.

Appendix F

Capacitor Color Codes

Not all capacitors are color coded. Virtually all filter capacitors, whether encased in aluminum housing or in a cardboard casing, will have values and voltage ratings imprinted on them. For smaller coupling or bypass capacitors the imprinting is also used as a convenient reference when replacement is required. Identifying color codes have, however, been used on the common ceramic capacitors illustrated in Fig. F-1.

Tubular capacitors may have *axial* leads, as shown in Fig. F-1(a), where the leads extend out from the ends. The leads may also be connected to the sides, as shown in Fig. F-1(b), in which case the connection is termed *radial*. The values identified by the digit color bands are in picofarads (pF). Another band rates the temperature coefficient, and an additional band indicates the tolerance. The temperature coefficient for ceramic capacitors is given in parts per million per degree Celsius (ppm/°C). When the letter N precedes the coefficient, a decrease in capacitance with an increase in temperature during operation is indicated. A positive-temperature coefficient is identified by the letter P, and a negative-positive-zero coefficient is indicated by the letters NPO. Thus, a designation of N220 indicates a capacitance decrease with a rise in temperature of 220 ppm/°C. The latter shows by how much the capacitance changes during the warm-up time of the unit in which the capacitor is used. Capacitors with an NPO rating are stable and have a negligible temperature effect on the capacitance value.

As indicated in Fig. F-1, five identification markings are generally used for these capacitors. For the axial lead capacitor, identification starts at the end where the color bands are grouped, with the leftmost

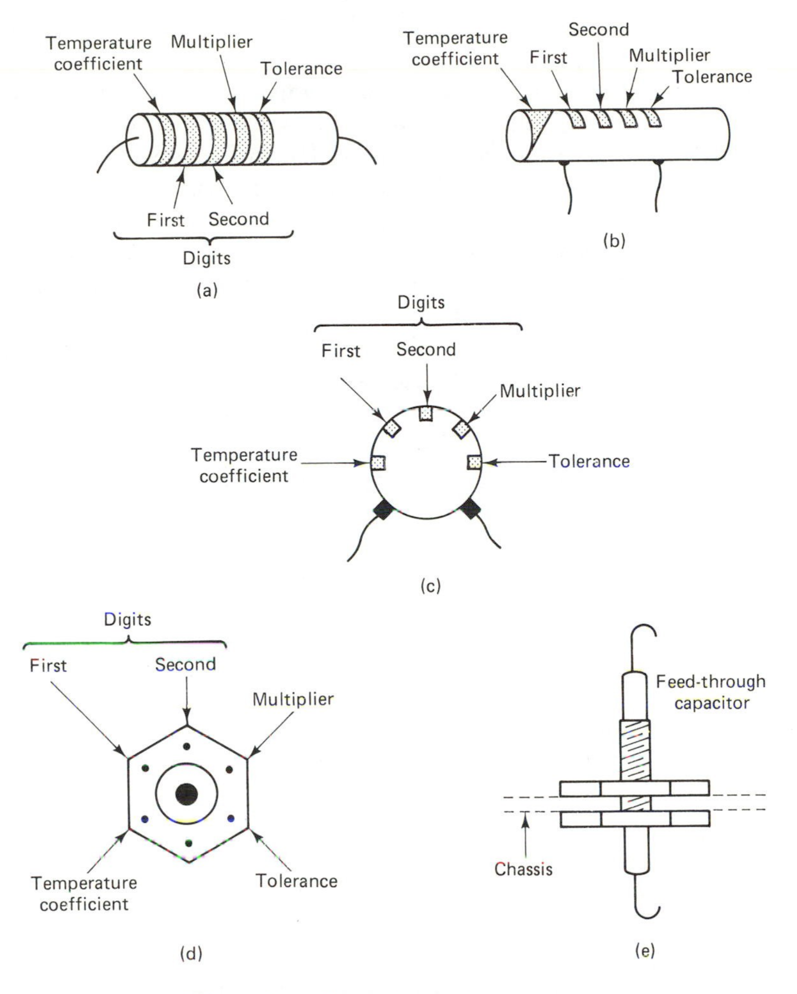

Figure F-1 Ceramic (ac-dc) capacitors.

band identifying the temperature coefficient. The next two bands indicate the significant digits making up the capacitance value. Often axial lead capacitors have a wide color marking for the first band. On occasion the radial lead capacitors have a wider initial band for easier identification. For the five-dot disk capacitor shown in Fig. F-1(c), the lower-left color dot indicates the temperature coefficient, and the other dots (in clockwise sequence) follow the same coding as the axial or radial lead capacitors. For the capacitor shown in Fig. F-1(c), the color markings are on one side only.

The feed-through ceramic capacitor is shown in both Fig. F-1(d) and (e). The color coding is in clockwise sequence, starting with the dot without color coding. The representation in Fig. F-1(e) is a side view of the feed-through capacitor. A threaded section is present, permitting rigid mounting by use of the bolt, as shown. On occasion the capacitor shown in Fig. F-1(b) comes in an extended-range-temperature-coefficient type. The color coding for the latter utilizes five color dots, with the first dot representing the temperature coefficient multiplier instead of a digit. In this instance the last four dots are the same as those shown in Fig. F-1(b).

Table F-1 shows the color coding for the five-dot capacitor, the extended-range capacitor, and the disk type. For the last there are occasions when only three color dots are used, in which case the temperature coefficient and tolerance values are not given. For the three-dot types the first two dots (reading clockwise) are the significant digits, and the last dot is the multiplier.

TABLE F-1 Capacitor Color Codes

				Five-Dot Temp. Coeff., TC	Extended Range	
Color	Digit	Multiplier	Tolerance		Significant Digits	Multiplier
Black	0	1	20%	NPO	0.0	−1
Brown	1	10	1%	N033		−10
Red	2	100	2%	N075	1.0	−100
Orange	3	1000	3%	N150	1.5	−1000
Yellow	4	10,000		N220	2.2	−10,000
Green	5		5%	N330	3.3	+1
Blue	6			N470	4.7	+10
Violet	7			N750	7.5	+100
Gray	8	0.01 (alt.)		*a*	*b*	+1000
White	9	0.1 (alt.)	10%	*c*		+10,000
Silver		0.01 (pref.)				
Gold		0.1 (pref.)				

[a]General-purpose types with a TC ranging from P150 to N1500.

[b]Coupling, decoupling, and general bypass types with a TC ranging from P100 to N750.

[c]If the first band (TC) is black, the range is N1000 to N5000.

Index

CANADORE COLLEGE NORTH BAY
WITHDRAWN